Linear Algebra and Geometry

Linear Algebra and Geometry

James A. Murtha
Earl R. Willard
Marietta College

Holt, Rinehart and Winston, Inc.
New York Chicago San Francisco Atlanta
Dallas Montreal Toronto London Sydney

Library of Congress Catalog Card Number: 69–20456
SBN: 03–074485–7
AMS 1968 Subject Classifications 1501, 5020, 5030
Printed in the United States of America
1 2 3 4 5 6 7 8 9

Preface

One of the common criticisms of the college mathematics curriculum is the extent to which it ignores geometry. The lack of emphasis on geometry was brought to the attention of the authors several years ago when we were asked to teach a course in geometry in an Academic Year Institute at Rutgers University for high-school teachers seeking a Master's degree in mathematics. We realized then that our own background in geometry was limited and concentrated in the historical aspect of the subject. We studied geometry, decided what we wanted to teach, and wrote notes for the students. This book has been developed from those notes.

The entire book reflects the interest of the authors in algebra. Chapter 1 develops the basic theory of vector spaces. In Chapters 2 and 3 an ample amount of geometry is done in an algebraic setting. Chapter 4 is essentially a response to a question frequently raised by students; this being that although geometry can be done in the algebraic setting of vector spaces, why should it appear as a sequence of corollaries to algebraic results? We answer this question by showing that suitable geometric axioms lead to vector spaces and that, with suitable modification, the statement "geometry and linear algebra are the same subject" can be made into a theorem.

Although this material was originally written with high-school teachers in mind, it has been modified with the hope that it will be of use to other students who wish to learn something about the relationship of classical geometry to modern mathematics. The book contains enough material for a two-semester course for the able student. For the class wishing to proceed at a more leisurely pace, or with less time to devote to the subject, it might be pointed out that Chapter 4, except for the last section on conics, is independent of Chapter 3. Thus it is possible to give a course ignoring the metric aspects of geometry by covering Chapters 1, 2, and 4. Students with a good background in linear algebra could use Chapters 2 and 4 to study geometry. Chapters 1 and 3 are, in essence, a linear algebra course covering, with the exception of canonical forms, the material generally covered in a linear algebra course at the advanced undergraduate or beginning graduate level.

One feature of the book which merits comment is the central role played by the exercises. We believe it important that the students actively participate in the development of the theory. With this in mind, we have included many exercises which are called upon in later developments. To each such exercise we have attached an asterisk along with a notation such as [2.3], indicating that the result is first referrred to in Chapter 2, Section 2.3.

We are indebted to the many people who aided us along the way. To Hiram Paley, John de Pillis, and especially to Victor Klee, we express our appreciation for their encouraging comments and criticism regarding the manuscript. We thank Miss Lauri Ross for typing the classroom notes during the first year we taught this material. Anne Murtha typed the final manuscript and various revisions; she deserves special thanks for her continual encouragement. We also want to thank Miss Emilie Bogart for her assistance in the final preparation stages of the manuscript. We are grateful to the able and personable staff of Holt, Rinehart and Winston, Inc., for their cooperation in preparing this book. Above all, we acknowledge the many students and colleagues who have forced us to clarify our thinking and, hopefully, our exposition.

James A. Murtha

Earl R. Willard

Marietta, Ohio

February 1969

Contents

Linear Algebra and Geometry

Introduction

Although it has been the intent of the authors to write a "self-contained" account of certain aspects of linear algebra and geometry, the reader is expected to be familiar with some basic concepts of mathematics. This introduction is offered to set forth the prerequisites for understanding the material that follows as well as to establish some relevant terminology and notation.

We assume the reader knows the rudiments of set theory, in particular the definitions and properties of *intersection*, *union*, *Cartesian product*, and *inclusion* and the meaning of such symbols as $\forall$, $\exists$, $\{x : x^2 = 2\}$, and $\varnothing$. We also expect the reader to be familiar with the basic properties of the integers $\mathbb{Z}$, the rational number $\mathbb{Q}$, the real numbers $\mathbb{R}$, and the complex numbers $\mathbb{C}$, say to the extent that these occur in a calculus course. The principle of mathematical induction will be called upon from time to time in proofs, and we trust that the reader will understand this technique.

We assume further that the reader has seen enough analytic geometry so that our motivational remarks dealing with 2- and 3-dimensional vectors and how they are added and multiplied by a scalar indeed serve their purpose rather than appear as unknown phenomena.

Of paramount importance throughout the text is the notion of a *function* from one set to another. We write $f:S \to T$ to indicate "f is a function from a set S to a set T," and we thereby mean that f is a rule of correspondence which assigns to each element s in S some unique element t in T; the element t is called the *image of s under f* and we write $t = f(s)$, or quite frequently $t = (s)f$. In fact the reader should become accustomed to using $f(s)$ or $(s)f$ without prejudice. We sometimes call functions *maps* or *mappings*.

Recall that the set S on which the function f is defined is called the *domain of f*, and the set $\{(s)f : s \in S\}$, or simply $(S)f$, is called the *image of f*. If it happens that $(S)f = T$, we say that f sends (or maps) S *onto* T, or simply that $f: S \to T$ is *onto*. Of course it may happen that $(s_1)f = (s_2)f$ for distinct elements s_1 and s_2 of S. If, however, $(s_1)f = (s_2)f$ implies $s_1 = s_2$, then we say that f is *one-to-one*, and write f is 1-1.

If $f: S \to T$ and $g:T \to W$ are functions, we can define the *composite* function $fg:S \to W$ by $(s)fg = ((s)f)g$ for $s \in S$. Note that if we are writing $f(s)$ and $g(t)$ rather than $(s)f$ and $(t)g$, then the composite function would be written gf. (In either case we think of "f followed by g.")

In case $f:S \to T$ is both 1-1 and onto, we can define the *inverse function* $f^{-1}:T \to S$ by

$$(t)f^{-1} = s \quad \text{if } (s)f = t.$$

In this event we have, of course, that ff^{-1} and $f^{-1}f$ are respectively the *identity functions* on S and T, that is,

$$(s)ff^{-1} = s, \quad \forall s \in S, \quad \text{and} \quad (t)f^{-1}f = t, \quad \forall t \in T.$$

A particular kind of function which occurs frequently is a *binary operation* on a set $S \neq \varnothing$, which is a function

$$g:S \times S \to S$$

whose domain is the Cartesian product of S with itself and whose range is a subset of S. Well-known examples of binary operations are addition and multiplication on $\mathbb{R}$, and composition of functions on the set S of all functions from a set T into itself.

EXERCISES

Note that the statements in the following problems are true whether the functions are written on the left or on the right.

1. Let $S = \{a_1, a_2, a_3\}$.

   (a) Make a list of all functions from S into itself.

(b) Among the functions from S into S are those which are 1-1 and onto. Make a list of these and show that if f and g are such functions, then so are fg and f^{-1}.

(c) In general, for any set S, a 1-1 function from S onto S is called a *permutation of S*. Show that the product and the inverse of permutations are again permutations.

2. Let $A = \{0, 1\}$, and let $S = A \times A$. Make a list of the permutations of S.

3. Suppose $f: S \to T$ and $g: T \to S$ are such that fg and gf are the identity maps on their respective domains. Show that f and g are 1-1 and onto, and that $g = f^{-1}, f = g^{-1}$.

Chapter 1 Finite Dimensional Vector Spaces

SECTION 1.1 VECTOR SPACES

Before making a formal definition of a vector space, let us consider a few examples.

When you studied analytic geometry you identified the Euclidean plane with $\mathbb{R}^2 = \{(x, y) : x \in \mathbb{R}, y \in \mathbb{R}\}$. Given two distinct points $(x_1, y_1) \in \mathbb{R}^2$, $(x_2, y_2) \in \mathbb{R}^2$ you learned that the line joining these two points has for its equation

$$(y - y_1)(x_2 - x_1) = (x - x_1)(y_2 - y_1).$$

More precisely, the line L joining the two points consists of the set of points satisfying the equation, that is,

$$L = \{(x, y) \in \mathbb{R}^2 : (y - y_1)(x_2 - x_1) = (x - x_1)(y_2 - y_1)\}.$$

An alternative description of L is motivated by parametric equations. Suppose $t \in \mathbb{R}$ and set $x = x_1 + t(x_2 - x_1)$, $y = y_1 + t(y_2 - y_1)$. Then

$$\begin{aligned}(y - y_1)(x_2 - x_1) &= t(y_2 - y_1)(x_2 - x_1)\\ &= t(x_2 - x_1)(y_2 - y_1)\\ &= (x - x_1)(y_2 - y_1),\end{aligned}$$

so

$$(x_1 + t(x_2 - x_1), y_1 + t(y_2 - y_1)) \in L.$$

Conversely, suppose $(x, y) \in L$. Since (x_1, y_1) and (x_2, y_2) are distinct points either $x_1 \neq x_2$ or $y_1 \neq y_2$ (or both). If $x_1 \neq x_2$ we set $t = (x - x_1)(x_2 - x_1)^{-1}$ and obtain

$$\begin{aligned}x_1 + t(x_2 - x_1) &= x_1 + (x - x_1)(x_2 - x_1)^{-1}(x_2 - x_1)\\ &= x_1 + x - x_1 = x,\end{aligned}$$

as well as

$$\begin{aligned}y_1 + t(y_2 - y_1) &= y_1 + (x - x_1)(x_2 - x_1)^{-1}(y_2 - y_1)\\ &= y_1 + (x - x_1)(y_2 - y_1)(x_2 - x_1)^{-1}\\ &= y_1 + (y - y_1)(x_2 - x_1)(x_2 - x_1)^{-1}\\ &= y_1 + y - y_1 = y.\end{aligned}$$

We have shown that

$$L = \{(x_1 + t(x_2 - x_1), y_1 + t(y_2 - y_1)) : t \in \mathbb{R}\},$$

provided $x_1 \neq x_2$. If $x_1 = x_2$, then $y_1 \neq y_2$. Set $t = (y - y_1)(y_2 - y_1)^{-1}$; after similar computations the same conclusion is reached.

This description of L lends itself naturally to the use of vectors. If we identify the point (x, y) with the vector, or directed line segment, from $(0, 0)$ to (x, y) the parallelogram rule of addition of plane vectors yields

$$(x_1, y_1) + (x_2, y_2) = (x_1 + x_2, y_1 + y_2).$$

(See Figure 1.1.) If t is a positive real number, the product of t with the vector (x, y) is defined to be a vector whose direction is the same as (x, y) and whose length is t times the length of (x, y). If t is a negative real number, its product with the vector (x, y) is defined to be a vector in the opposite direction to (x, y) and of length $-t$ times the length of (x, y). In either case, $t \cdot (x, y) = (tx, ty)$.

The line L passing through the points (x_1, y_1) and (x_2, y_2) can be described in terms of the addition of vectors and the multiplication of a vector by a real number by

$$\begin{aligned}L &= \{(x_1, y_1) + t[(x_2, y_2) - (x_1, y_1)] : t \in \mathbb{R}\}\\ &= \{(1 - t)(x_1, y_1) + t(x_2, y_2) : t \in \mathbb{R}\}.\end{aligned}$$

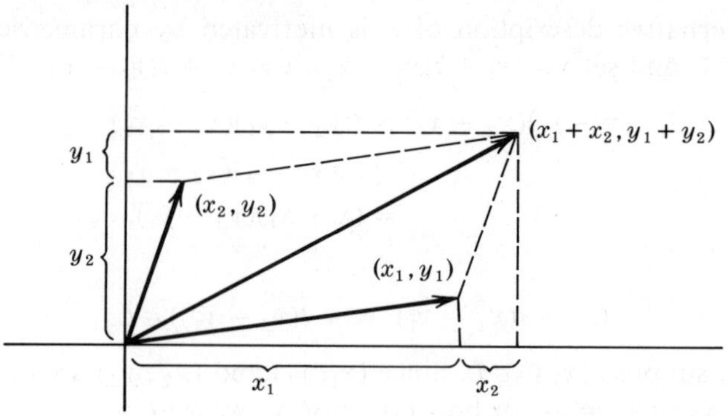

Figure 1.1

A similar discussion can be made in connection with Euclidean space. The points are identified with $\mathbb{R}^3 = \{(x, y, z): x \in \mathbb{R}, y \in R, z \in \mathbb{R}\}$, and $(x, y, z) \in \mathbb{R}^3$ is identified with the vector from (0, 0, 0) to (x, y, z). Addition of vectors is given by $(x_1, y_1, z_1) + (x_2, y_2, z_2) = (x_1 + x_2, y_1 + y_2, z_1 + z_2)$, and multiplication of a vector by a real number is given by $t(x, y, z) = (tx, ty, tz)$. Suitable adaptation of the formulas you learned in analytic geometry leads to the description of the line L joining distinct points (x_1, y_1, z_1) and (x_2, y_2, z_2) by $L = \{(1 - t)(x_1, y_1, z_1) + t(x_2, y_2, z_2): t \in \mathbb{R}\}$, and the plane P determined by the distinct points (x_1, y_1, z_1), (x_2, y_2, z_2), and (x_3, y_3, z_3) by

$$P = \{(1 - s - t)(x_1, y_1, z_1) + s(x_2, y_2, z_2) + t(x_3, y_3, z_3): s \in \mathbb{R}, t \in \mathbb{R}\}.$$

Indeed, given a set V, the elements of which we will call vectors or points, together with a definition of the addition of points and a definition of scalar multiplication, that is, the product of a real number and a point, it is possible to define lines and planes. We can define the line joining the distinct points p_1 and p_2 in V to be $\{(1 - t)p_1 + tp_2: t \in \mathbb{R}\}$ and the plane determined by the noncollinear points $p_1, p_2, p_3 \in V$ to be $\{(1 - s - t)p_1 + sp_2 + tp_3: s, t \in \mathbb{R}\}$. If the addition and scalar multiplication have reasonable properties we can anticipate that these points, lines, and planes will behave in much the way that our experience with plane and solid Euclidean geometry leads us to expect.

The reader may agree that such an abstraction is feasible, but he may maintain substantial doubt as to its value. If this is the initial encounter with abstract axiomatic structures, such skepticism is understandable, but we ask a temporary suspension of judgment and hope the developments in this book will be a convincing argument for the value of this approach to the study of geometry.

One reason for the central role the study of abstract axiomatic structures has achieved in modern mathematics is the fact that once a theorem has been proven in such a context, the theorem is valid in every concrete situation where the axioms are satisfied.

For example, consider the familiar geometric theorem stating that the medians of a triangle intersect in a point. In the plane the proof goes as follows. Let (x_1, y_1), (x_2, y_2), (x_3, y_3) be the vertices of a triangle. The midpoint of the side joining (x_2, y_2) and (x_3, y_3) is

$$\left(\frac{x_2 + x_3}{2}, \frac{y_2 + y_3}{2}\right) = \frac{1}{2}(x_2, y_2) + \frac{1}{2}(x_3, y_3),$$

and the median from the vertex (x_1, y_1) is

$$\{(1 - t)(x_1, y_1) + t(\tfrac{1}{2}(x_2, y_2) + \tfrac{1}{2}(x_3, y_3)) : t \in \mathbb{R}\}.$$

The median from (x_2, y_2) is

$$\left\{(1 - t)(x_2, y_2) + \frac{t}{2}(x_1, y_1) + \frac{t}{2}(x_3, y_3) : t \in \mathbb{R}\right\}.$$

Similarly, the median from (x_3, y_3) is

$$\left\{(1 - t)(x_3, y_3) + \frac{t}{2}(x_1, y_1) + \frac{t}{2}(x_2, y_2) : t \in \mathbb{R}\right\}.$$

Letting $t = \frac{2}{3}$ in each case, one sees that the point

$$\tfrac{1}{3}(x_1, y_1) + \tfrac{1}{3}(x_2, y_2) + \tfrac{1}{3}(x_3, y_3)$$

is on each of the three lines. (To complete the proof, one must show that the medians do not have another point in common; this fact follows from the observation that two points determine a line, and that since the original vertices are not collinear, the medians must be distinct.)

If we want to prove the same theorem in Euclidean space, the basic idea of the proof is the same but the notation has to be changed. The three vertices are points in space, (x_1, y_1, z_1), (x_2, y_2, z_2), and (x_3, y_3, z_3). The midpoint of the side joining (x_2, y_2, z_2) and (x_3, y_3, z_3) is

$$\left(\frac{x_2 + x_3}{2}, \frac{y_2 + y_3}{2}, \frac{z_2 + z_3}{2}\right) = +\frac{1}{2}(x_2, y_2, z_2) + \frac{1}{2}(x_3, y_3, z_3)$$

and the median through (x_1, y_1, z_1) is

$$\{(1 - t)(x_1, y_1, z_1) + t(\tfrac{1}{2}(x_2, y_2, z_2) + \tfrac{1}{2}(x_3, y_3, z_3)) : t \in \mathbb{R}\}$$

or

$$\left\{(1 - t)(x_1, y_1, z_1) + \frac{t}{2}(x_2, y_2, z_2) + \frac{t}{2}(x_3, y_3, z_3) : t \in \mathbb{R}\right\}.$$

The other medians are

$$\left\{(1-t)(x_2, y_2, z_2) + \frac{t}{2}(x_1, y_1, z_1) + \frac{t}{2}(x_3, y_3, z_3) : t \in \mathbb{R}\right\}$$

and

$$\left\{(1-t)(x_3, y_3, z_3) + \frac{t}{2}(x_1, y_1, z_1) + \frac{t}{2}(x_2, y_2, z_2) : t \in \mathbb{R}\right\}.$$

One then observes that the point

$$\tfrac{1}{3}(x_1, y_1, z_1) + \tfrac{1}{3}(x_2, y_2, z_2) + \tfrac{1}{3}(x_3, y_3, z_3)$$

is common to all three lines.

If we merely postulate a set of points together with an addition of points and a scalar multiplication satisfying a few simple arithmetic properties, we can provide a similar proof that the medians of a triangle intersect at a point. This proof then serves to prove the theorem in both the Euclidean plane and space, since the postulates are satisfied in each case. It also serves to prove the theorem in a variety of other contexts; we will soon want to consider these.

We do not, however, wish to claim that the study of "vector spaces" (the formal name for a set of points together with a suitably axiomatized addition and scalar multiplication) is completely justified by the fact that their use enables us to prove theorems in both the Euclidean plane and space at the same time. Any such economy is more than offset by the work involved in developing the theory of vector spaces sufficiently to do a significant amount of geometry.

The point is not simply that vector spaces provide a context in which geometry can be done but, roughly speaking, that geometric considerations lead naturally to vector spaces. More exactly, we will prove in Chapter 4 that, given a rather modest set of axioms about points and lines, there is a vector space such that every theorem following from the geometric axioms is also a theorem in the vector space and vice versa. In a sense, the study of geometry and the study of vector spaces are equivalent. This statement needs important qualification, but one of the major aims of this book is to prove the properly qualified version of the equivalence. We will then have at least two ways to prove a given statement: the type of proof found in high-school geometry, a so-called synthetic proof, and a proof using the algebraic techniques of vector spaces.

There is, however, a difficulty that cannot be ignored. If we wish to study vector spaces in such a way as to prove the aforementioned equivalence, we cannot restrict our attention to the situation in which scalar multiplication is necessarily the product of a real number and a vector. We are forced to introduce a more general kind of number and define scalar multiplication to be the product of one of these numbers with a vector. Unfortunately, when (as in Definition 1) we write down the axioms for these numbers (or scalars as they are called), few if any of you will know of any concrete examples that satisfy these axioms and do not satisfy the commutative law of multiplication, that is, $\alpha\beta = \beta\alpha$. Such examples are complicated; we will not provide you with one until Section 1.8. In the meantime you should feel free to think of the scalars as real numbers, noticing, however, that we use only the properties listed in Definition 1.

DEFINITION 1 A *vector space* consists of:

(a) A set V, the elements of which, are called *vectors* (or *points*), together with a binary operation $V \times V \to V$ called *vector addition* where for $x, y \in V$ the image of $(x, y) \in V \times V$, is denoted by $x + y$. Vector addition satisfies the following axioms:
- **(i)** for $x, y, z \in V$, $x + (y + z) = (x + y) + z$ (associative law);
- **(ii)** there exists $O \in V$ such that $x + O = x$ for all $x \in V$ (O is called an *identity* for V);
- **(iii)** for each $x \in V$ there exists $-x \in V$ such that $x + (-x) = O$ ($-x$ is called an *inverse of* x);
- **(iv)** for $x, y \in V$, $x + y = y + x$ (commutative law).

(b) A set D, the elements of which are called *scalars*, together with two binary operations $D \times D \to D$, one called *addition of scalars*, where the image of $(\alpha, \beta) \in D \times D$ is denoted $\alpha + \beta$, and the other called *multiplication of scalars*, where the image of $(\alpha, \beta) \in D \times D$ is denoted $\alpha\beta$. These operations satisfy the following axioms:
- **(i)** for $\alpha, \beta, \gamma \in D$, $\alpha + (\beta + \gamma) = (\alpha + \beta) + \gamma$ (associative law of addition) and $\alpha(\beta\gamma) = (\alpha\beta)\gamma$ (associative law of multiplication);
- **(ii)** there exist distinct elements $0, 1 \in D$ such that for all $\alpha \in D$, $\alpha + 0 = \alpha$ and $\alpha \cdot 1 = 1 \cdot \alpha = \alpha$ (0 is called an *additive identity* of D and 1 is called a *multiplicative identity* of D);
- **(iii)** for each $\alpha \in D$ there exists $-\alpha \in D$ such that $\alpha + (-\alpha) = 0$ ($-\alpha$ is called an *additive inverse* of α) and, if $\alpha \neq 0$, there exists $\alpha^{-1} \in D$ such that $\alpha\alpha^{-1} = \alpha^{-1}\alpha = 1$ (α^{-1} is called a *multiplicative inverse* of α);
- **(iv)** for $\alpha, \beta \in D$, $\alpha + \beta = \beta + \alpha$ (commutative law of addition);
- **(v)** for $\alpha, \beta, \gamma \in D$, $\alpha(\beta + \gamma) = \alpha\beta + \alpha\gamma$ and $(\beta + \gamma)\alpha = \beta\alpha + \gamma\alpha$ (distributive laws).

(c) A map called *scalar multiplication* $D \times V \to V$, where the image of $(\alpha, x) \in D \times V$ is denoted by αx, satisfying the following axioms:
- **(i)** for $\alpha, \beta \in D$, $x \in V$, $(\alpha\beta)x = \alpha(\beta x)$ (associative law of scalar multiplication);
- **(ii)** for $x \in V$, $1 \cdot x = x$ (unitary law);
- **(iii)** for $\alpha, \beta \in D$, $x, y \in V$, $(\alpha + \beta)x = \alpha x + \beta x$ and $\alpha(x + y) = \alpha x + \alpha y$ (distributive laws).

We shall denote the vector space with V as its set of vectors and D as its scalars by ${}_DV$.

Any set D satisfying the axioms listed in part (b) is called a *division ring*. The real number system $\mathbb{R}$, together with the usual definitions of addition and multiplication, is an example of a division ring. The set $\mathbb{R}$, of course, enjoys additional properties. Not only do we have $\alpha \cdot \beta = \beta \cdot \alpha$, $\forall \alpha, \beta \in \mathbb{R}$, but we also have present in $\mathbb{R}$ a notion of "order." Given $\alpha, \beta \in \mathbb{R}$, we can say $\alpha < \beta$, $\alpha = \beta$, or $\beta < \alpha$. Division rings in general do not have any such order relation. Other familiar examples of division rings are $\mathbb{C}$, the complex numbers, and $\mathbb{Q}$, the rational numbers, with, in each case, the usual definitions of addition and multiplication.

The following examples give some idea of the variety of vector spaces that occur in mathematics.

EXAMPLE 1 Let D be a division ring and let $D^n = \{(\alpha_1, \cdots, \alpha_n) : \alpha_i \in D, i = 1, \cdots, n\}$. We define, for $(\alpha_1, \cdots, \alpha_n), (\beta_1, \cdots, \beta_n) \in D^n$,

$$(\alpha_1, \cdots, \alpha_n) + (\beta_1, \cdots, \beta_n) = (\alpha_1 + \beta_1, \cdots, \alpha_n + \beta_n).$$

Since $\alpha_i, \beta_i \in D \Rightarrow \alpha_i + \beta_i \in D$ we have $(\alpha_1 + \beta_1, \cdots, \alpha_n + \beta_n) \in D^n$, and so "$+$" is a binary operation $D^n \times D^n \to D^n$. It satisfies the axioms to make D^n into a set of vectors. For example, if

$$(\alpha_1, \cdots, \alpha_n), (\beta_1, \cdots, \beta_n), (\gamma_1, \cdots, \gamma_n) \in D^n,$$

then

$$\begin{aligned}
((\alpha_1, \cdots, \alpha_n) &+ (\beta_1, \cdots, \beta_n)) + (\gamma_1, \cdots, \gamma_n) \\
&= (\alpha_1 + \beta_1, \cdots, \alpha_n + \beta_n) + (\gamma_1, \cdots, \gamma_n) \\
&= ((\alpha_1 + \beta_1) + \gamma_1, \cdots, (\alpha_n + \beta_n) + \gamma_n) \\
&= (\alpha_1 + (\beta_1 + \gamma_1), \cdots, \alpha_n + (\beta_n + \gamma_n)) \\
&= (\alpha_1, \cdots, \alpha_n) + ((\beta_1 + \gamma_1), \cdots, (\beta_n + \gamma_n)) \\
&= (\alpha_1, \cdots, \alpha_n) + ((\beta_1, \cdots, \beta_n) + (\gamma_1, \cdots, \gamma_n)),
\end{aligned}$$

and so the associative law of addition holds. Then $(0, \cdots, 0)$ serves as the additive identity of D^n since

$$(\alpha_1, \cdots, \alpha_n) + (0, \cdots, 0) = (\alpha_1 + 0, \cdots, \alpha_n + 0) = (\alpha_1, \cdots, \alpha_n).$$

The additive inverse of $(\alpha_1, \cdots, \alpha_n)$ is $(-\alpha_1, \cdots, -\alpha_n)$ since

$$(\alpha_1, \cdots, \alpha_n) + (-\alpha_1, \cdots, -\alpha_n) = (\alpha_1 + (-\alpha_1), \cdots, \alpha_n + (-\alpha_n)) = (0, \cdots, 0).$$

The addition is commutative since

$$\begin{aligned}
(\alpha_1, \cdots, \alpha_n) + (\beta_1, \cdots, \beta_n) &= (\alpha_1 + \beta_1, \cdots, \alpha_n + \beta_n) \\
&= (\beta_1 + \alpha_1, \cdots, \beta_n + \alpha_n) \\
&= (\beta_1, \cdots, \beta_n) + (\alpha_1, \cdots, \alpha_n).
\end{aligned}$$

The division ring D of scalars is given, so there is no need to check part (b) of Definition 1. If we define scalar multiplication by

$$\alpha(\alpha_1, \cdots, \alpha_n) = (\alpha\alpha_1, \cdots, \alpha\alpha_n),$$

then

$$\begin{aligned}(\alpha\beta)(\alpha_1, \cdots, \alpha_n) &= ((\alpha\beta)\alpha_1, \cdots, (\alpha\beta)\alpha_n)\\ &= (\alpha(\beta\alpha_1), \cdots, \alpha(\beta\alpha_n))\\ &= \alpha(\beta\alpha_1, \cdots, \beta\alpha_n)\\ &= \alpha(\beta(\alpha_1, \cdots, \alpha_n)).\end{aligned}$$

Moreover,

$$\begin{aligned}1 \cdot (\alpha_1, \cdots, \alpha_n) &= (1 \cdot \alpha_1, \cdots, 1 \cdot \alpha_n) = (\alpha_1, \cdots, \alpha_n),\\ (\alpha + \beta)(\alpha_1, \cdots, \alpha_n) &= ((\alpha + \beta)\alpha_1, \cdots, (\alpha + \beta)\alpha_n)\\ &= (\alpha\alpha_1 + \beta\alpha_1, \cdots, \alpha\alpha_n + \beta\alpha_n)\\ &= (\alpha\alpha_1, \cdots, \alpha\alpha_n) + (\beta\alpha_1, \cdots, \beta\alpha_n)\\ &= \alpha(\alpha_1, \cdots, \alpha_n) + \beta(\alpha_1, \cdots, \alpha_n),\end{aligned}$$

and finally,

$$\begin{aligned}\alpha((\alpha_1, \cdots, \alpha_n) + (\beta_1, \cdots, \beta_n)) &= \alpha(\alpha_1 + \beta_1, \cdots, \alpha_n + \beta_n)\\ &= (\alpha(\alpha_1 + \beta_1), \cdots, \alpha(\alpha_n + \beta_n))\\ &= (\alpha\alpha_1 + \alpha\beta_1, \cdots, \alpha\alpha_n + \alpha\beta_n)\\ &= (\alpha\alpha_1, \cdots, \alpha\alpha_n) + (\alpha\beta_1, \cdots, \alpha\beta_n)\\ &= \alpha(\alpha_1, \cdots, \alpha_n) + \alpha(\beta_1, \cdots, \beta_n),\end{aligned}$$

and so ${}_D D^n$ satisfies the axioms for a vector space. Whenever ${}_D D^n$ is written we will assume that it consists of the set of n-tuples with entries from D together with the operations as defined above. See Exercises 5 and 6 for alternative operations on the set of n-tuples.

The proof that the following examples are vector spaces will be left as an exercise for the reader.

EXAMPLE 2 Let S be a set and $F(S)$ the set of all functions $f{:}S \to \mathbb{R}$. If $f, g \in F(S)$, define $f + g{:}S \to \mathbb{R}$ by $(f + g)(s) = f(s) + g(s)$, $\forall s \in S$, and for $\alpha \in \mathbb{R}$ define $\alpha f{:}S \to \mathbb{R}$ by $(\alpha f)(s) = \alpha(f(s))$. ${}_{\mathbb{R}}F(S)$ is a vector space.

In particular let $S = [0, 1] = \{t \in \mathbb{R}{:}0 \leq t \leq 1\}$. Then $C(S) = \{f \in F(S){:}f$ is continuous$\}$ and $D(S) = \{f \in F(S){:}f$ is differentiable$\}$ are vector spaces under the same definition of addition and scalar multiplication used in $F(S)$.

EXAMPLE 3 Let $\mathbb{R}[x]$ denote the set of polynomials in the indeterminate x with real coefficients. ${}_{\mathbb{R}}\mathbb{R}[x]$ under the usual addition of polynomials and multiplication by a constant is a vector space.

EXAMPLE 4 Letting $\mathbb{C}$ = complex numbers, $\mathbb{R}$ = real numbers, and $\mathbb{Q}$ = rational numbers, ${}_{\mathbb{Q}}\mathbb{R}$, ${}_{\mathbb{Q}}\mathbb{C}$, ${}_{\mathbb{R}}\mathbb{C}$ are vector spaces.

We shall now prove a number of elementary properties of vector spaces.

Proposition 1

Let ${}_DV$ be a vector space.

(a) The identity of V is unique;
(b) For each $v \in V$, the inverse of v is unique;
(c) The additive and multiplicative identities of D are unique;
(d) The additive inverse of $\alpha \in D$ is unique, and, if $\alpha \neq 0$, its multiplicative inverse is unique.

Proof

(a) Suppose we have a $z \in V$ such that $x + z = x$ for all $x \in V$. Then $z = z + O = O + z = O$. Note that the first equality comes from the fact that O is an identity, the second from the commutative law, and the third from the assumption that z is an additive identity. The proofs of both parts of part (c) are similar to the proof of part (a).
(b) By axioms we have $-v$ such that $v + (-v) = O$. Suppose we have a $z \in V$ such that $v + z = O$. Then

$$\begin{aligned} z = z + O &= z + (v + (-v)) \\ &= (z + v) + (-v) \\ &= (v + z) + (-v) \\ &= O + (-v) \\ &= -v. \end{aligned}$$

The proofs of both parts of part (d) are similar to the proof of part (a).

Proposition 2

Let $\alpha \in D$, $x \in V$. Then $\alpha x = O \Leftrightarrow \alpha = 0$ or $x = O$.

Proof

$\boxed{\Leftarrow}$: $0 \cdot x = (0 + 0)x = 0 \cdot x + 0 \cdot x$. Therefore

$$\begin{aligned} O &= 0 \cdot x + (-(0 \cdot x)) = (0 \cdot x + 0 \cdot x) + (-(0 \cdot x)) \\ &= 0 \cdot x + (0 \cdot x + -(0 \cdot x)) = 0 \cdot x + O = 0 \cdot x. \end{aligned}$$

Similarly, $\alpha \cdot O = \alpha(O + O) = \alpha \cdot O + \alpha \cdot O$, and so

$$O = \alpha \cdot O + (-(\alpha \cdot O)) = (\alpha \cdot O + \alpha \cdot O) + (-(\alpha \cdot O))$$
$$= \alpha \cdot O + (\alpha \cdot O + (-(\alpha \cdot O))) = \alpha \cdot O + O = \alpha \cdot O.$$

$\boxed{\Rightarrow}$: Suppose $\alpha \cdot x = O$. If $\alpha = 0$ we are done. If $\alpha \neq 0$, α^{-1} exists and

$$O = \alpha^{-1} \cdot O = \alpha^{-1}(\alpha x) = (\alpha^{-1}\alpha)x = 1 \cdot x = x.$$

Proposition 3

If $\alpha \in D$, $x \in V$, then $-(\alpha x) = (-\alpha)x$.

Proof

$\alpha x + (-\alpha)x = (\alpha + (-\alpha))x = 0 \cdot x = O$, and so by the uniqueness of the inverse (Proposition 1) we have $(-\alpha)x = -(\alpha x)$.

Corollary

If $x \in V$, then $-x = (-1)x$.

Proof

Let $\alpha = 1$ in Proposition 3.

Proposition 4

Let ${}_DV$ be a vector space.

(a) If $x \in V$, then $-(-x) = x$.
(b) If $\alpha \in D$, then $-(-\alpha) = \alpha$.
(c) If $\alpha \in D$, $\alpha \neq 0$, then $(\alpha^{-1})^{-1} = \alpha$.

Proof

Part (a): $x + (-x) = (-x) + x = O$, and so by the definition and uniqueness of inverses $-(-x) = x$. Parts (b) and (c) have similar proofs.

Corollary

Let ${}_DV$ be a vector space.

(a) If $x, y \in V$, then $-x = -y \Leftrightarrow x = y$.
(b) If $\alpha, \beta \in D$, then $-\alpha = -\beta \Leftrightarrow \alpha = \beta$.
(c) If $\alpha, \beta \in D$, $\alpha \neq 0$, $\beta \neq 0$, then $\alpha^{-1} = \beta^{-1} \Leftrightarrow \alpha = \beta$.

Proof

The proof is immediate from Proposition 3 and the uniqueness of inverses.

Roughly speaking, what we have shown with these propositions is that the various symbols 0, 1, +, −, · as we have defined them behave in the same way as they have since you first encountered them in arithmetic as children. The following proposition, though simple in statement and proof, is an important fact about vector spaces.

Proposition 5

Let $x \in V$, $x \neq O$, and $\alpha, \beta \in D$. Then $\alpha x = \beta x \Leftrightarrow \alpha = \beta$.

Proof

$\boxed{\Leftarrow}$: The proof is obvious.

$$\boxed{\Rightarrow}: \alpha x = \beta x \Rightarrow O = \alpha x + (-(\beta x)) = \alpha x + (-\beta)x$$
$$= (\alpha + (-\beta))x \Rightarrow \alpha + (-\beta) = 0$$
$$\Rightarrow -\alpha = -\beta \Rightarrow \alpha = \beta.$$

As a simplification of notation we will write $x - y$ for $x + (-y)$, $x, y \in V$ and $\alpha - \beta$ for $\alpha + (-\beta)$, $\alpha, \beta \in D$.

EXERCISES

1. Show that the examples in Examples 2, 3, and 4 of the text are vector spaces.

2. Let D be a division ring, $\alpha, \beta \in D$, $\alpha \neq 0$, $\beta \neq 0$. Show that $(\alpha\beta)^{-1} = \beta^{-1}\alpha^{-1}$.

3. Let $f \in F([0, 1])$. (See Example 2 for notation.) Show that $\{y \in D([0, 1]): (dy/dt) + f(t) \cdot y(t) = 0\}$ is a vector space.

4. Show that the set of all (x, y, z) that are solutions of $2x - 3y + z = 0$ is a vector space where the vectors are some elements of $\mathbb{R}^3$.

5. Let $V = \{(\alpha, \beta, \gamma): \alpha, \beta, \gamma \in \mathbb{R}\}$. Define $(\alpha_1, \beta_1, \gamma_1) + (\alpha_2, \beta_2, \gamma_2) = (\alpha_1 + \alpha_2, \beta_1 + \beta_2, 0)$ and $\delta(\alpha, \beta, \gamma) = (\delta\alpha, \delta\beta, \delta\gamma)$. Show that with these definitions of addition and scalar multiplication V is not a vector space.

6. Let $V = \{(\alpha, \beta):\alpha, \beta \in \mathbb{C}\}$. For all $\alpha, \alpha_1, \alpha_2, \beta_1, \beta_2 \in \mathbb{C}$, define $(\alpha_1, \beta_1) + (\alpha_2, \beta_2) = (\alpha_1 + \alpha_2, \beta_1 + \beta_2)$ and $\alpha(\alpha_1, \beta_1) = (\bar{\alpha}\alpha_1, \bar{\alpha}\beta_1)$, where $\bar{\alpha}$ denotes the complex conjugate of α. Show that V is a vector space.

7. Show that the cancellation law holds for the addition of vectors, that is, let $u, v, w \in V$, a vector space; if $u + v = u + w$, then $v = w$.

SECTION 1.2

SUBSPACES

In the vector space $\mathbb{R}^2$, the subset $L_1 = \{(\alpha, 0):\alpha \in \mathbb{R}\}$, that is the x-axis, is in its own right a vector space over $\mathbb{R}$; the same definitions of addition and scalar multiplication are used as given for ${}_{\mathbb{R}}\mathbb{R}^2$. For example,

$$\begin{aligned} (\alpha, 0) + (\beta, 0) &= (\alpha + \beta, 0) \in L_1, \forall \alpha, \beta, \\ \beta(\alpha, 0) &= (\beta\alpha, 0) \in L_1, \forall \alpha, \beta, \\ -(\alpha, 0) &= (-\alpha, 0) \in L_1, \forall \alpha, \end{aligned}$$

and so on, the other axioms being easily verified. The y-axis $L_2 = \{(0, \beta):\beta \in \mathbb{R}\}$ is also a vector space using these same operations.

More generally, any other line L passing through the origin is given by $\{(\alpha, \beta):\beta = \mu\alpha$, for some fixed $\mu \in \mathbb{R}\}$, where μ is the slope of L. Once again, direct computation shows that L is a vector space using the operations of ${}_{\mathbb{R}}\mathbb{R}^2$. These simple examples motivate:

DEFINITION 1 Let ${}_DV$ be a vector space. A subset $L \subseteq V, L \neq \varnothing$, is called a *subspace* of V if ${}_DL$ is a vector space under the same operations of vector addition and scalar multiplication as those in V. We denote the fact that L is a subspace of V by $L < V$.

Remark: To appreciate the notion of a subspace, one should consider how a subset of a vector space could fail to be a subspace. Suppose L is a subset of V. For $u, v \in L$, the element $u + v$ is in V but need not be an element of L. However, if L is to be a subspace of V, $u + v$ must be in L. Similarly for $\alpha \in D$ and $u \in L$, αu is an element of V, and, if L is to be a subspace, αu must be in L.

More formally, addition in V is a binary operation, that is, a *function*

$$\sigma:V \times V \to V \text{ where } \sigma(u, v) = u + v.$$

We are interested in the *restriction* of this function to the domain $L \times L$; in particular we want to know whether σ maps $L \times L$ *into* L. Similarly, we want to know whether the function

$$\tau : D \times V \to V \text{ given by } \tau(\alpha, u) = \alpha u,$$

when restricted to $D \times L$ has its image in L.

An analogous but more familiar problem concerns the set of integers and the usual binary operation, addition. The subset of even integers is distinguished from the subset of odd integers in that addition is a binary operation on the former but not on the latter.

In each case it makes sense to talk about the addition of two elements in the subset, that is, the restriction of the addition function is meaningful. Yet in one case the image is contained in the subset but in the other case it is not.

Of course to say that a subset L of a vector space V is a vector space in its own right is saying much more than that addition and scalar multiplication are binary operations on L. Nevertheless, as Proposition 1 shows, the remaining properties are inherited from the structure of V.

EXAMPLE 1 For any vector space ${}_DV$, $\{O\} < V$ and $V < V$. A common abuse of notation is to use simply O instead of $\{O\}$ to denote the subspace consisting of the element O.

EXAMPLE 2 In terms of the examples of vector spaces given in Section 1.1 we have

$$D([0, 1]) < C([0, 1]) < F([0, 1]).$$

If ${}_DV$ is given with $L \subseteq V$, then in order to determine whether ${}_DL$ is a subspace of V it is necessary only to check parts (a) and (c) of the definition of a vector space; part (b) is of course automatically satisfied. The following proposition gives us simpler criteria for the determination.

Proposition 1

Let ${}_DV$ be a vector space and $L \subseteq V$, $L \neq \varnothing$. Then the following statements are equivalent:

(a) $L < V$;
(b) $\alpha \in D$ and $u, v \in L \Rightarrow u + v \in L$ and $\alpha u \in L$;
(c) $\alpha, \beta \in D$ and $u, v \in L \Rightarrow \alpha u + \beta v \in L$.

Proof

(a) $\Rightarrow$ **(b)** $L < V$ says that L is a vector space under the same binary operations as V, so that the maps $V \times V \to V$ and $D \times V \to V$ when restricted from V to L give maps $L \times L \to L$ and $D \times L \to L$. Hence $u + v \in L$ and $\alpha u \in L$.

(b) ⇒ (c) By (b), $\alpha u \in L$ and $\beta v \in L$, and hence $\alpha u + \beta v \in L$.

(c) ⇒ (a) Let $u, v \in L$. Then $1 \cdot u + 1 \cdot v = u + v \in L$, so $V \times V \to V$ restricts to $L \times L \to L$. If $u, v, w \in L$, then $u, v, w \in V$, so $u + (v + w) = (u + v) + w$. Since $L \neq \varnothing$, there is an element $u \in L$, and $O = 0 \cdot u + 0 \cdot u \in L$. If $u \in L$, then $(-1) \cdot u + 0 \cdot u = -u \in L$. If $u, v \in L$, then $u, v \in V$, so $u + v = v + u$, and all the axioms for vector addition are satisfied.

If $u \in L, \alpha \in D$, then $\alpha u + 0 \cdot u = \alpha u \in L$, so the map $D \times V \to V$ restricts to $D \times L \to L$. If $\alpha, \beta \in D$, $u, v \in L$, then $u, v \in V$, so $(\alpha\beta)u = \alpha(\beta u)$, $1 \cdot u = u$, $(\alpha + \beta)u = \alpha u + \beta u$, and $\alpha(u + v) = \alpha u + \alpha v$, and all the properties of scalar multiplication are verified. Hence $L < V$.

A useful extension of the equivalence of (a) and (c) in the proposition is readily available provided we understand what is meant by the sum of n elements of V.

DEFINITION 2 Let ${}_D V$ be a vector space. The sum of two elements is defined in the axioms for a vector space and, provided the sum of $n - 1$ elements of V has been defined, we define the sum of $u_1, \cdots, u_n \in V$ by $u_1 + \cdots + u_n = (u_1 + \cdots + u_{n-1}) + u_n$.

Proposition 2

Let ${}_D V$ be a vector space, $u_1, \cdots, u_n \in V$, and k an integer, $1 \leq k < n$. Then $(u_1 + \cdots + u_k) + (u_{k+1} + \cdots + u_n) = u_1 + \cdots + u_n$.

Proof

The proof is by induction on n, the case $n = 3$ being the associative law in the axioms for V. Suppose that for any $m < n$ and integer t, $1 \leq t < m$ and any $v_1, \cdots, v_m \in V$, we know that $(v_1 + \cdots + v_t) + (v_{t+1} + \cdots + v_m) = v_1 + \cdots + v_m$.

Consider $(u_1 + \cdots + u_k) + (u_{k+1} + \cdots + u_n)$, If $k = n - 1$ the conclusion of the proposition is the definition. If $k < n - 1$ then $u_{k+1} + \cdots + u_n = (u_{k+1} + \cdots + u_{n-1}) + u_n$ by definition and

$$
\begin{aligned}
(u_1 + \cdots + u_k) &+ (u_{k+1} + \cdots + u_n) \\
&= (u_1 + \cdots + u_k) + ((u_{k+1} + \cdots + u_{n-1}) + u_n) \\
&= ((u_1 + \cdots + u_k) + (u_{k+1} + \cdots + u_{n-1})) + u_n \\
&= (u_1 + \cdots + u_{n-1}) + u_n \\
&= u_1 + \cdots + u_n .
\end{aligned}
$$

Proposition 3

If $L < V$, then $\alpha_1 u_1 + \alpha_2 u_2 + \cdots + \alpha_n u_n \in L$ for any $\{\alpha_1, \alpha_2, \cdots, \alpha_n\} \subseteq D$, and for any $\{u_1, \cdots, u_n\} \subseteq L$, and conversely.

Proof

Let $L < V$. By Proposition 1, the assertion is true when $n = 1$. Suppose it is true for $n - 1$ terms, and that we are given $\{\alpha_1, \cdots, \alpha_n\} \subseteq D$ and $\{u_1, \cdots, u_n\} \subseteq L$. Then

$$\begin{aligned} \alpha_1 u_1 + \alpha_2 u_2 + \cdots + \alpha_n u_n &= (\alpha_1 u_1 + \cdots + \alpha_{n-1} u_{n-1}) + \alpha_n u_n \\ &= u' + \alpha_n u_n \in L, \end{aligned}$$

since, by inductive hypothesis, $u' \in L$. The converse is obvious.

The next definition gives us methods of deriving new subspaces from old.

DEFINITION 3 Let ${}_D V$ be a vector space with $L < V$ and $M < V$. The *sum* of L and M and the *intersection* of L and M are, respectively,

$$\begin{aligned} L + M &= \{x + y \in V : x \in L \text{ and } y \in M\}; \\ L \cap M &= \{x \in V : x \in L \text{ and } x \in M\}. \end{aligned}$$

Proposition 4

Let ${}_D V$ be a vector space with $L < V$ and $M < V$. Then $L + M < V$ and $L \cap M < V$.

Proof

Let $x + y \in L + M$ and $x' + y' \in L + M$ with $x, x' \in L$, and $y, y' \in M$. Let $\alpha, \beta \in D$. Then

$$\begin{aligned} \alpha(x + y) + \beta(x' + y') &= \alpha x + \alpha y + \beta x' + \beta y' \\ &= (\alpha x + \beta x') + (\alpha y + \beta y') \in L + M. \end{aligned}$$

Hence $L + M < V$. If $x, y \in L \cap M$, and $\alpha, \beta \in D$, then $\alpha x + \beta y \in L$ and $\alpha x + \beta y \in M$; hence $\alpha x + \beta y \in L \cap M$, so that $L \cap M < V$.

Proposition 5

Let ${}_D V$ be a vector space, N any set, and for each $\nu \in N$, let L_ν be a subspace of V. Put

$$L = \bigcap_{\nu \in N} L_\nu = \{x \in V : x \in L_\nu \text{ for all } \nu \in N\}.$$

Then $L < V$.

Proof

Let $x, y \in L$. Then $x, y \in L_\nu$ for each $\nu \in N$. Let $\alpha, \beta \in D$. Then $\alpha x + \beta y \in L_\nu$, for each $\nu \in N$, so that $\alpha x + \beta y \in L$. Hence $L < V$.

EXERCISES

1. Let $U, W < V$, where V is any vector space. Show
   (a) $U + U = U \cap U = U$
   (b) $U + W = W + U$
   (c) $U < U + W$, and $W < U + W$
   (d) if $L < V$ with $U < L$ and $W < L$, then $U + W < L$.

2. Let $x \in {}_D V$, $x \neq O$. Put $L = \{\alpha x : \alpha \in D\}$. Show
   (a) $L < V$, and
   (b) if $M < L$, then $M = O$ or $M = L$.

3. Which of the following subsets of ${}_{\mathbb{R}}\mathbb{R}^3$ are subspaces?
   (a) $\{(\alpha, \beta, \gamma) : \gamma = 0\}$.
   (b) $\{(\alpha, \beta, \gamma) : \alpha + \beta + \gamma = 0\}$.
   (c) $\{(\alpha, \beta, \gamma) : \alpha^2 + \beta^2 = \gamma\}$.
   (d) $\{(\alpha, \beta, \gamma) : \alpha + \beta + \gamma = 2\}$.

*4. Let (α_1, α_2) and (β_1, β_2) be nonzero elements of ${}_{\mathbb{R}}\mathbb{R}^2$ such that $(\alpha_1, a_2) \notin \{\mu(\beta_1, \beta_2) : \mu \in \mathbb{R}\}$.
   (a) Show that $\alpha_1\beta_2 - \alpha_2\beta_1 \neq 0$. [*Hint*: Consider two cases, $\alpha_1 \neq 0$ and $\alpha_2 \neq 0$; by hypothesis at least one is true.]
   (b) Let $(\gamma_1, \gamma_2) \in \mathbb{R}^2$. Show that there exists $\alpha, \beta \in \mathbb{R}$ such that $\alpha(\alpha_1, \alpha_2) + \beta(\beta_1, \beta_2) = (\gamma_1, \gamma_2)$. [*Hint*: Rewrite this vector equation as a system of two linear equations.]
   (c) Let $L_1 = \{\alpha(\alpha_1, \alpha_2) : \alpha \in \mathbb{R}\}$, and $L_2 = \{\beta(\beta_1, \beta_2) : \beta \in \mathbb{R}\}$. Show that $L_1 + L_2 = \mathbb{R}^2$. [1.3]

5. (a) Let $L < \mathbb{R}^2$, $L \neq O$, $L \neq \mathbb{R}^2$. Show that there exists $(\alpha, \beta) \in \mathbb{R}^2$ such that $L = \{\mu(\alpha, \beta) : \mu \in \mathbb{R}\}$.
   (b) Let $L, M < \mathbb{R}^2$, $L \neq O$, $M \neq O$, $L \neq M$. Show $L + M = \mathbb{R}^2$.

SECTION 1.3
GENERATORS AND LINEARLY INDEPENDENT SETS

The vectors (1, 0) and (0, 1) in $\mathbb{R}^2$ have the property that if $(\alpha, \beta) \in \mathbb{R}^2$, then $(\alpha, \beta) = \alpha(1, 0) + \beta(0, 1)$; that is, any element of $\mathbb{R}^2$ can be written as a sum of scalar multiples of (1, 0) and (0, 1). Other sets of vectors have this property, for example (1, 1) and (1, −1), since $(\alpha, \beta) = [(\alpha + \beta)/2](1, 1) + [(\alpha - \beta)/2]\ (1, -1)$ for all $(\alpha, \beta) \in \mathbb{R}^2$. A third example is the set (−1, −1), (−1, 1), (0, 1), since $(\alpha, \beta) = -(\alpha/2)(-1, -1) - (\alpha/2)(-1, 1) + \beta(0, 1)$ for all $(\alpha, \beta) \in \mathbb{R}^2$.

The last example differs from the others in at least two ways. Although the third set has three elements, each of the others has only two. A second difference is with respect to a uniqueness property. Specifically, $(\alpha, \beta) = \alpha(1, 0) + \beta(0, 1)$ is the only way (α, β) can be represented as a sum of scalar multiples of (1, 0) and (0, 1) for

$$(\alpha, \beta) = \lambda(1, 0) + \mu(0, 1) \Rightarrow \alpha = \lambda \quad \text{and} \quad \beta = \mu.$$

A similar uniqueness statement holds for the pair (1, 1), (1, −1); but with the set (−1, −1), (−1, 1), (0, 1) we have not only

$$(\alpha, \beta) = -\frac{\alpha}{2}(-1, -1) - \frac{\alpha}{2}(-1, 1) + \beta(0, 1),$$

but also

$$(\alpha, \beta) = \alpha(-1, -1) - 2\alpha(-1, 1) + (3\alpha + \beta)(0, 1).$$

The fundamental ideas underlying these simple observations are studied in this section.

DEFINITION 1 Let ${}_DV$ be a vector space, $x_1, \cdots, x_n \in V$, $\alpha_1, \cdots, \alpha_n \in D$. The element $\alpha_1 x_1 + \cdots + \alpha_n x_n \in V$ is called a *linear combination* of $x_1, \cdots, x_n$. We shall frequently use the notation $\sum_{i=1}^{n} \alpha_i x_i$ for $\alpha_1 x_1 + \cdots + \alpha_n x_n$.

DEFINITION 2 Let S be a nonempty subset of the vector space ${}_DV$. We denote by $\langle S \rangle$ the set of all linear combinations of finite subsets of S. $\langle S \rangle$ is called the *span* of S and S is said to *span* or *generate* $\langle S \rangle$.

The remarks at the beginning of this section showed that

$$\mathbb{R}^2 = \langle\{(1, 0), (0, 1)\}\rangle = \langle\{(1, 1), (1, -1)\}\rangle = \langle\{(-1, -1), (-1, 1), (0, 1)\}\rangle.$$

If x is an indeterminate and $S = \{1, x, x^2, \cdots, x^n, \cdots\}$, then $\langle S \rangle = \mathbb{R}[x]$. This example shows the significance of taking only linear combinations of finite subsets of S, for if we took infinite linear combinations of S we would have power series rather than polynomials. If ${}_DV$ is a vector space then $\langle V \rangle = V$, and if $x \in V$ then $\langle \{x\} \rangle = \{\alpha x : \alpha \in D\}$. We shall usually write $\langle x \rangle$ for $\langle \{x\} \rangle$ and $\langle x_1, \cdots, x_n \rangle$ for $\langle \{x_1, \cdots, x_n\} \rangle$.

Proposition 1

Let ${}_DV$ be a vector space, $S \subseteq V$, $S \neq \varnothing$. Then:

(a) $\langle S \rangle < V$;

(b) $S = \langle S \rangle \Leftrightarrow S < V$.

Proof

(a) Let $x_1, \cdots, x_n \in S$, $y_1, \cdots, y_m \in S$, $\alpha_1, \cdots, \alpha_n$, $\beta_1, \cdots, \beta_m \in D$, so that $\sum_{i=1}^n \alpha_i x_i$ and $\sum_{i=1}^m \beta_i y_i \in \langle S \rangle$. If $\alpha, \beta \in D$ then

$$\alpha\left(\sum_{i=1}^{n} \alpha_i x_i\right) + \beta\left(\sum_{i=1}^{m} \beta_i y_i\right)$$
$$= \alpha\alpha_1 x_1 + \cdots + \alpha\alpha_n x_n + \beta\beta_1 y_1 + \cdots + \beta\beta_m y_m \in \langle S \rangle \Rightarrow \langle S \rangle$$

is a subspace of V.

(b) $\boxed{\Rightarrow}$: The proof is immediate from (a).

$\boxed{\Leftarrow}$: If $x_1, \cdots, x_n \in S$, $\alpha_1, \cdots, \alpha_n \in D$ so that $\alpha_1 x_1 + \cdots + \alpha_n x_n \in \langle S \rangle$, the fact that $S < V$ implies $\alpha_1 x_1 + \cdots + \alpha_n x_n \in S$, and so $\langle S \rangle \subseteq S$. Since $S \subseteq \langle S \rangle$ we have $S = \langle S \rangle$.

Proposition 2

Let ${}_DV$ be a vector space, $S \subseteq V$, $S \neq \varnothing$, $L < V$, and $S \subseteq L$. Then $\langle S \rangle < L$.

Proof

The proof is left to the reader.

Proposition 3

Let ${}_DV$ be a vector space, $S \subseteq V$, $S \neq \varnothing$. Let $\mathcal{S} = \{L < V : S \subseteq L\}$. Then $\langle S \rangle = \bigcap_{L \in \mathcal{S}} L$.

Proof

$\langle S \rangle \in \mathcal{S} \Rightarrow \bigcap_{L \in \mathcal{S}} L \subseteq \langle S \rangle$. Conversely, if $L \in \mathcal{S}$ then $\langle S \rangle \subseteq L \Rightarrow \langle S \rangle \subseteq \bigcap_{L \in \mathcal{S}} L$.

This characterization of $\langle S \rangle$ could well be taken as a definition. It says that $\langle S \rangle$ is the smallest subspace of V containing S.

DEFINITION 3 Let ${}_DV$ be a vector space. A set $\{x_1, \cdots, x_n\} \subseteq V$ is said to be *linearly independent* if, for $\alpha_1, \cdots, \alpha_n \in D$

$$\alpha_1 x_1 + \cdots + \alpha_n x_n = O \Rightarrow \alpha_1 = \cdots = \alpha_n = 0.$$

An infinite subset S of V is said to be linearly independent if every finite subset of S is linearly independent. A subset S of V that is not linearly independent is said to be *linearly dependent*.

If the reader will check the three examples of spanning sets for $\mathbb{R}^2$ given at the beginning of this section, he will find that the first two sets are linearly independent and the third linearly dependent.

Proposition 4

Let ${}_DV$ be a vector space.

(a) Any subset S of V containing O is linearly dependent.
(b) If $x \in V$, $x \neq O$, then $\{x\}$ is linearly independent.
(c) If $S \subseteq S' \subseteq V$ and S' is linearly independent then so is S.
(d) If $S \subseteq S' \subseteq V$ and S is linearly dependent then so is S'.

Proof

The proof is left to the reader.

Proposition 5

Let $\varnothing \neq S \subseteq {}_DV$, $\langle S \rangle \neq O$. Then S is linearly dependent $\Leftrightarrow \exists x \in S$ such that $x \in \langle S \backslash \{x\} \rangle$.

Proof

$\boxed{\Rightarrow}$: If $O \in S$ then $O \in \langle S \backslash \{O\} \rangle$, since $S \backslash \{O\} \neq \varnothing$. Otherwise if S is linearly dependent we have $x_1, \cdots, x_n \in S$, and $\alpha_1, \cdots, \alpha_n \in D$, with

$$\alpha_1 x_1 + \cdots + \alpha_n x_n = O,$$

and one of the scalars, $\alpha_i \neq 0$. Then $x_i = \sum_{j=1, j \neq i}^{n} \alpha_i^{-1} \alpha_j x_j$, where at least one $\alpha_j \neq 0$ since $x_i \neq O$, and hence $x_i \in \langle S \backslash \{x_i\} \rangle$.

$\boxed{\Leftarrow}$: If $x \in \langle S \backslash \{x\} \rangle$ we have $x_1, \cdots, x_n \in S$, $x_i \neq x$ for all i and $\alpha_1, \cdots, \alpha_n \in D$ such that $x = \alpha_1 x_1 + \cdots + \alpha_n x_n \Rightarrow O = -x + \alpha_1 x_1 + \cdots + \alpha_n x_n \Rightarrow S$ linearly dependent.

Corollary

Let S be a nonempty subset of V. S is linearly dependent $\Leftrightarrow \exists S' \subsetneq S$ such that $\langle S' \rangle = \langle S \rangle$.

DEFINITION 4 Let ${}_DV$ be a vector space and $S \subseteq V$. S is called a *basis* of V if

(a) $\langle S \rangle = V$;
(b) S is linearly independent.

We know that $\mathbb{R}^2$ has a basis, $\{(1, 0), (0, 1)\}$. A natural question is: Do all vector spaces have a basis? The answer is Yes, but the proof, in full generality, depends upon a fairly sophisticated knowledge of set theory. We shall content ourselves with a more special result that will be sufficient for the applications of vector spaces to geometry.

DEFINITION 5 Let ${}_DV$ be a vector space. We say V is *finitely generated* if there exists a finite set $S \subseteq V$ such that $\langle S \rangle = V$.

Theorem 1

Every finitely generated nonzero vector space has a basis.

Proof

Suppose ${}_DV = \langle S \rangle$ and S is finite. If S is linearly independent then S is a basis. If not there exists $S_1 \subsetneqq S$ such that $\langle S_1 \rangle = \langle S \rangle = V$. If S_1 is linearly independent it is a basis; if not there exists $S_2 \subsetneqq S_1$ such that $\langle S_2 \rangle = \langle S_1 \rangle = V$. Since S was finite, only a finite number of elements can be removed, and therefore after a finite number of deletions we must obtain a basis.

We conclude this section with a characterization of linearly independent sets that gives bases much of their utility.

Theorem 2

Let ${}_DV$ be a vector space and $S = \{x_1, \cdots, x_n\} \subseteq V$. S is linearly independent $\Leftrightarrow$ every $x \in \langle S \rangle$ has a unique representation $x = \alpha_1 x_1 + \cdots + \alpha_n x_n$, $\alpha_i \in D$.

Proof

$\boxed{\Rightarrow}$: Suppose $x = \alpha_1 x_1 + \cdots + \alpha_n x_n = \beta_1 x_1 + \cdots + \beta_n x_n$. Then $O = (\alpha_1 - \beta_1)x_1 + \cdots + (\alpha_n - \beta_n)x_n$, so if S is linearly independent $\alpha_i - \beta_i = 0$ for all i.

$\boxed{\Leftarrow}$: If $\alpha_1 x_1 + \cdots + \alpha_n x_n = O$ then, since $O = 0 \cdot x_1 + \cdots + 0 \cdot x_n$ and representations are unique, we have $\alpha_i = 0$ for all i, and hence S is linearly independent.

EXERCISES

1. Let $S = \{(2, 1), (1, -1), (-1, -1)\} \subseteq \mathbb{R}^2$. Show that there exists a subset $S' \subsetneq S$ such that $\langle S' \rangle = \langle S \rangle$.

2. Generalize Exercise 1 to show that any three vectors in $\mathbb{R}^2$ are linearly dependent. [*Hint*: Use Exercise 4, Section 1.2.]

3. Is it true that if $x, y, z \in {}_D V$ are linearly independent, then $\{x + y, y + z, z + x\}$ is linearly independent?

4. Under what conditions on $\alpha \in \mathbb{R}$ is the set $\{(\alpha, 1, 0), (1, \alpha, 1), (0, 1, \alpha)\}$ linearly independent in $\mathbb{R}^3$?

5. Under what conditions on $\alpha \in \mathbb{R}$ is the set $\{(1 + \alpha, 1 - \alpha), (1 - \alpha, 1 + \alpha)\}$ linearly independent in $\mathbb{R}^2$?

*6. Let ${}_D V$ be a vector space and $S = \{x_1, \cdots, x_n\}$, n distinct elements of V. Let $\mathscr{D}(S) = \{(\alpha_1, \cdots, \alpha_n) \in D^n : \alpha_1 x_1 + \cdots + \alpha_n x_n = O\}$. Show that $\mathscr{D}(S) < D^n$ and that S is linearly independent $\Leftrightarrow \mathscr{D}(S) = O$. [3.2]

7. Let $V = D^n$. Let $e_i \in D^n$, $1 \leq i \leq n$ be the vector with a 1 in the ith position and 0 elsewhere. Then show that $\{e_1, \cdots, e_n\}$ is a basis for D^n. We shall often refer to this as the *standard basis of* D^n.

8. Let $\{x_1, \cdots, x_n\}$ be a linearly independent subset of ${}_D V$. Show that the set $\{x_1, x_1 + x_2, x_1 + x_3, \cdots, x_1 + x_n\}$ is also linearly independent.

*9. Let ${}_D V$ be a vector space and $L_1, \cdots, L_n$ subspaces of V. We define

$$L_1 + \cdots + L_n = \{x_1 + \cdots + x_n : x_i \in L_i, i = 1, \cdots, n\}.$$

Let $\mathscr{S} = \{M < V : L_i \subseteq M, i = 1, \cdots, n\}$. Prove:
(a) $L_1 + \cdots + L_n = \langle L_1 \cup \cdots \cup L_n \rangle$;
(b) $L_1 + \cdots + L_n = \bigcap_{M \in \mathscr{S}} M$.
Note that, as a consequence of part (a) or part (b), $L_1 + \cdots + L_n$ is a subspace of V. [1.4]

SECTION 1 4

DIMENSION AND DIRECT SUMS

Throughout this section let ${}_DV$ be an arbitrary but fixed finitely generated vector space. We know by Theorem 1, Section 1.3, that V has a basis and, from the example of $\mathbb{R}^2$, that it may have more than one. We shall now show that any two bases have the same number of elements.

Proposition 1

Let $S = \{x_1, \cdots, x_n\}$, $T = \{y_1, \cdots, y_m\} \subseteq V$, S linearly independent and $\langle T \rangle \subseteq \langle S \rangle$. Then $m > n \Rightarrow T$ is linearly dependent.

Proof

The proof is by induction on n. If $n = 1$ and $m > 1$ then $m \geq 2$. If $y_1 = 0$ or $y_2 = 0$ we are done. If not, $y_1 = \alpha_1 x_1$, $y_2 = \alpha_2 x_1$ with $\alpha_1 \neq 0$, $\alpha_2 \neq 0$. Therefore $\alpha_1^{-1} y_1 - \alpha_2^{-1} y_2 = 0$ and T is linearly dependent. Now assume that if S is any set of $n - 1$ linearly independent vectors in V, the result holds. Then $\langle T \rangle \subseteq \langle S \rangle \Rightarrow y_i = z_i + \alpha_i x_n$ for $i = 1, \cdots, m$ with $z_i \in \langle x_1, \ldots, x_{n-1} \rangle$. If $\alpha_i = 0$ for all i then $y_i \in \langle x_1, \cdots, x_{n-1} \rangle$, $\forall i$, so $\langle T \rangle \subseteq \langle x_1, \cdots, x_{n-1} \rangle$ and $m > n \Rightarrow m > n - 1$ so $\{y_1, \cdots, y_m\}$ is linearly dependent by the induction hypothesis. If not, $\alpha_i \neq 0$ for some i, say $\alpha_j \neq 0$, $1 \leq j \leq m$. Therefore $x_n = \alpha_j^{-1}(y_j - z_j)$ and for each i,

$$y_i = z_i + \alpha_i \alpha_j^{-1}(y_j - z_j) \Rightarrow y_i - \alpha_i \alpha_j^{-1} y_j \in \langle x_1, \cdots, x_{n-1} \rangle$$

for $i \neq j$, $1 \leq i \leq m$. This is a set of $m - 1$ vectors and $m - 1 > n - 1$ so they are linearly dependent, and hence there exist $\beta_1, \cdots, \beta_m \in D$ such that $\beta_1(y_1 - \alpha_1 \alpha_j^{-1} y_j) + \cdots + \beta_m(y_m - \alpha_m \alpha_j^{-1} y_j) = O$ and some $\beta_i \neq 0$. This yields

$$O = \beta_1 y_1 + \cdots + \beta_{j-1} y_{j-1} + \beta_{j+1} y_{j+1} + \cdots + \beta_m y_m - (\beta_1 \alpha_1 + \cdots + \beta_m \alpha_m)\alpha_j^{-1} y_j,$$

and hence T is linearly dependent.

Theorem 1

Let $S = \{x_1, \cdots, x_n\}$, $T = \{y_1, \cdots, y_m\}$ be bases of ${}_DV$. Then $n = m$.

Proof

If $m \neq n$, then $m > n$ or $n > m$. If $m > n$, then by Proposition 1, T is linearly dependent, which is contrary to hypothesis, and if $n > m$ Proposition 1 says S is linearly dependent, which is also false. Therefore $m = n$.

DEFINITION 1 The number of elements in a basis of a finitely generated vector space V is called the *dimension* of V and denoted $\dim(V)$. As a special case $\dim(O) = 0$. Finitely generated vector spaces are more commonly called *finite dimensional.*

Proposition 2

If $\dim({}_DV) = n$ and $\{x_1, \cdots, x_n) \subseteq V$ is linearly independent, then $\{x_1, \cdots, x_n\}$ is a basis for V.

Proof

If $x \in V$ and $x \notin \langle x_1, \cdots, x_n\rangle$, then the set $\{x_1, \cdots, x_n, x\}$ is linearly independent. But $\dim(V) = n$ implies there exists a basis $\{y_1, \cdots, y_n\}$ of V, and $\langle x_1, \cdots, x_n, x\rangle \subseteq \langle y_1, \cdots, y_n\rangle$ implies (by Proposition 1) that $\{x_1, \cdots, x_n, x\}$ is linearly dependent. Therefore $V = \langle x_1, \cdots, x_n\rangle$ and $\{x_1, \cdots, x_n\}$ is a basis for V.

In proving the existence of a basis we took a generating set for V and removed elements to obtain the basis. We now show that one can go in the other direction; that is, one can begin with a linearly independent set and extend it to a basis by adding elements.

Theorem 2

If $\dim(V) = n$, $m < n$, and $\{y_1, \cdots, y_m\}$ is a linearly independent set, then there exist $x_1, \cdots, x_{n-m} \in V$ such that $\{y_1, \cdots, y_m, x_1, \cdots, x_{n-m}\}$ is a basis.

Proof

If $m < n$ then $\langle y_1, \cdots, y_m\rangle \subsetneqq V$, so there exists $x_1 \in V$, with $x_1 \notin \langle y_1, \cdots, y_m\rangle$ such that $\{y_1, \cdots, y_m, x_1\}$ is linearly independent. If $m + 1 = n$ we are done. If $m + 1 < n$ there exists $x_2 \in V, x_2 \notin \langle y_1, \cdots, y_m, x_1\rangle$, and so $\{y_1, \cdots, y_m, x_1, x_2\}$ is linearly independent. After repeating this process $n - m$ times we arrive at a linearly independent set containing n elements which, by Proposition 2, is a basis.

Proposition 3

Let ${}_DV$ be a finite dimensional vector space and $L < V$. Then $\dim(L) \leq \dim(V)$ and $\dim(L) = \dim(V) \Leftrightarrow L = V$.

Proof

Note that a set of elements of L is linearly independent if and only if they are linearly independent when considered as elements of V; the result follows immediately from Proposition 2 and Theorem 2.

Corollary

Let $O \neq U_1 \subseteq U_2 \subseteq \cdots \subseteq U_m \subseteq V$, for subspaces U_i of V. If $\dim V < m$, then $U_i = U_{i+1}$, for at least one i.

Proof

Suppose, on the contrary, that $U_i \neq U_{i+1}$ for $i = 1, 2, \cdots, m-1$. Then $\dim U_1 \geq 1$, $\dim U_2 \geq 1 + \dim U_1 \geq 2, \cdots, \dim U_m \geq m$, a contradiction.

The following definition may, at first, seem unrelated to the discussion of bases and dimension. Much of the remainder of this section is devoted to making a connection.

DEFINITION 2 Let ${}_DV$ be a vector space, $L < V, M < V$. $L + M$ is called a *direct sum* and denoted $L \oplus M$ provided $L \cap M = O$. If $L \oplus M = V$, we say that L and M are *complements*, each of the other.

A few examples may clarify the situation. If $V = \mathbb{R}^2$ and L is any line through the origin, then from the earlier exercises any other line M through the origin is a complement of L. If $V = \mathbb{R}^3$ then the sum of any two distinct lines through the origin is direct; the sum of no two distinct planes through the origin is direct for they always intersect in a line. We are using in these comments some facts, such as that every nontrivial subspace of $\mathbb{R}^3$ is a line or plane through the origin, which will be proven by you in the exercises following this section. To continue with the discussion of examples, if L is a line through the origin in $\mathbb{R}^3$, then any plane M through the origin not containing L is a complement to L.

Proposition 4

Let ${}_DV$ be a vector space, $L < V$, $M < V$, and $V = L \oplus M$.

(a) If $x \in V$, then x has a unique representation $x = x_1 + x_2$ with $x_1 \in L$, $x_2 \in M$.
(b) If V is finite dimensional, then $\dim(L) + \dim(M) = \dim(V)$.

Proof

(a) $V = L \oplus M \Rightarrow V = L + M$, so if $x \in V$ there exist $x_1 \in L, x_2 \in M$ such that $x = x_1 + x_2$. Suppose we have $y_1 \in L, y_2 \in M$ such that $x = y_1 + y_2$. Then

$$O = x - x = (x_1 + x_2) - (y_1 + y_2) = (x_1 - y_1) + (x_2 - y_2).$$

Therefore $x_1 - y_1 = y_2 - x_2 \in L \cap M$. But $L \cap M = O$ by hypothesis and so $x_1 = y_1$ and $x_2 = y_2$.

(b) Let $\{x_1, \cdots, x_n\}$ be a basis for L and $\{y_1, \cdots, y_m\}$ a basis for M. We claim that $\{x_1, \cdots, x_n, y_1, \cdots, y_m\}$ is a basis for V. If $z \in V$ then $z = z_1 + z_2$, $z_1 \in L, z_2 \in M$. $z_1 = \alpha_1 x_1 + \cdots + \alpha_n x_n$ since $\{x_1, \cdots, x_n\}$ is a basis for L, and since $\{y_1, \cdots, y_m\}$ is a basis for M, $z_2 = \beta_1 y_1 + \cdots + \beta_m y_m$. Therefore $z = \alpha_1 x_1 + \cdots + \alpha_n x_n + \beta_1 y_1 + \cdots + \beta_m y_m$ and $\langle x_1, \cdots, x_n, y_1, \cdots, y_m \rangle = V$. To show the linear independence, suppose we have $\alpha_1 x_1 + \cdots + \alpha_n x_n + \beta_1 y_1 + \cdots + \beta_m y_m = O$. Then

$$\begin{aligned} \alpha_1 x_1 + \cdots + \alpha_n x_n &= -\beta_1 y_1 - \cdots - \beta_m y_m \in L \cap M \\ &= O \Rightarrow \alpha_1 = \cdots = \alpha_n \\ &= \beta_1 = \cdots = \beta_m = 0. \end{aligned}$$

We now wish to extend the notion of direct sum from the direct sum of two subspaces to the direct sum of a finite number of subspaces. (See Section 1.3, Exercise 9, for the definition of the sum of a finite number of subspaces.) One possible definition would be that if $L_1, \cdots, L_m <_D V$, the sum $L_1 + \cdots + L_m$ is direct if, for $i \neq j$, $L_i \cap L_j = O$. A consequence of this definition, however, is that we have no analogue to Proposition 4. For example, if we take $V = \mathbb{R}^3$ and L_1, L_2, L_3 to be distinct 1-dimensional subspaces contained in a 2-dimensional subspace, we have that the intersection of any pair is O and yet $L_1 \subseteq L_2 + L_3$, so a vector $x_1 \in L_1$ can be written $x_1 = x_2 + x_3$ with $x_2 \in L_2, x_3 \in L_3$, that is, $x_1 + 0 \cdot x_2 + 0 \cdot x_3 = 0 \cdot x_1 + 1 \cdot x_2 + 1 \cdot x_3 \in L_1 + L_2 + L_3$ and the uniqueness of representation is contradicted. In order to obtain an analogue of Proposition 4 we proceed as follows.

DEFINITION 3 Let ${}_DV$ be a vector space, $L_1, \cdots, L_m < V$. The sum $L_1 + \cdots + L_m$ is called *direct* and is written $L_1 \oplus \cdots \oplus L_m$ provided $L_i \cap (\sum_{j=1, j \neq i}^{m} L_j) = O$ for $i = 1, \cdots, m$.

Proposition 5

Let ${}_DV$ be a vector space, $L_1, \cdots, L_m < V$ and $V = L_1 \oplus \cdots \oplus L_m$.

(a) If $x \in V$, then x has a unique representation $x = x_1 + \cdots + x_m$ with $x_i \in L_i$.

(**b**) If V is finite dimensional then

$$\dim(V) = \dim(L_1) + \cdots + \dim(L_m).$$

Proof

The proof is left as an exercise for the reader.

EXERCISES

1. (a) Extend (1, 0, 1), (1, 1, 0) to a basis of $\mathbb{R}^3$.
   (b) Extend (1, 0, 1, 0), (1, 1, 0, 0) to a basis of $\mathbb{R}^4$.

2. Prove that D^n has dimension n.

3. Prove that $\mathbb{R}[x]$ is not finitely generated.

4. Let $\mathbb{R}^n[x] = \{p(x) \in \mathbb{R}[x]: \text{degree of } p(x) \leq n\}$. Show that $\dim \mathbb{R}^n[x] = n + 1$.

5. Let ${}_DV$ be a vector space, $L_1, L_2, L_3, L_4 < V$. Show that the sum $L_1 + L_2 + L_3 + L_4$ is direct $\Leftrightarrow O = L_1 \cap L_2 = (L_1 + L_2) \cap L_3 = (L_1 + L_2 + L_3) \cap L_4$.

6. Supply a proof for Proposition 5.

*7. If ${}_DV$ is a vector space, $L < V$, $M < V$, then prove $\dim(L + M) + \dim(L \cap M) = \dim(L) + \dim(M)$. [*Hint*: Show that $L = L_1 \oplus (L \cap M)$, $M = M_1 \oplus (L \cap M)$ and hence $L + M = L_1 \oplus M_1 \oplus (L \cap M)$.] [2.2]

8. Prove that any finite dimensional vector space is the direct sum of 1-dimensional subspaces.

9. Let ${}_DV$ be a finite dimensional vector space, $L < V$, $M < V$, $\dim(L) = l$, $\dim(M) = m$. Find the maximum and minimum dimensions for $L \cap M$. Find the maximum and minimum dimensions for $L + M$.

10. Let ${}_DV$ be a vector space, $L < V$, $M < V$. If $\dim(V) = n$ and $\dim(L) = \dim(M) = n - 1$, then either $L = M$ or $\dim(L \cap M) = n - 2$.

SECTION 1.5
HOMOMORPHISMS

In this section we shift our point of view. To this point we have studied the properties of an individual vector space. Homomorphisms give us a way of relating different vector spaces and also give information about individual spaces.

DEFINITION 1 Let ${}_DV$, ${}_DW$ be vector spaces. A *homomorphism* from V to W is a map $f:V \to W$ such that:

(a) $(x + y)f = (x)f + (y)f$ for all $x, y \in V$;
(b) $(\alpha x)f = \alpha(x)f$ for all $x \in V$, $\alpha \in D$.

Homomorphisms are also called *linear maps* or *linear transformations*. Let us consider a few examples.

The map $f:\mathbb{R}^2 \to \mathbb{R}^3$ defined by $(\alpha, \beta)f = (\alpha, \beta, 0)$ satisfies both axioms for a homomorphism, since

$$\begin{aligned}((\alpha_1, \beta_1) + (\alpha_2, \beta_2))f &= (\alpha_1 + \alpha_2, \beta_1 + \beta_2)f \\ &= (\alpha_1 + \alpha_2, \beta_1 + \beta_2, 0) \\ &= (\alpha_1, \beta_1, 0) + (\alpha_2, \beta_2, 0) \\ &= (\alpha_1, \beta_1)f + (\alpha_2, \beta_2)f,\end{aligned}$$

and

$$\begin{aligned}(\lambda(\alpha, \beta))f &= (\lambda\alpha, \lambda\beta)f \\ &= (\lambda\alpha, \lambda\beta, 0) \\ &= \lambda(\alpha, \beta, 0) \\ &= \lambda(\alpha, \beta)f.\end{aligned}$$

Other examples, the verification of which are left to the reader, are the following:

(1) the map $g:\mathbb{R}^3 \to \mathbb{R}^3$ defined by $(\alpha, \beta, \gamma)g = (\alpha, \beta, 0)$;
(2) the map $h:\mathbb{R}^3 \to \mathbb{R}^3$ defined by $(\alpha, \beta, \gamma)h = (\alpha, \gamma, \beta)$;
(3) the differentiation map $d:D([0, 1]) \to F([0, 1])$, where $D([0, 1])$ is, you recall, the vector space of functions differentiable on the unit interval, and $F([0, 1])$ is the vector space of functions on the unit interval.

We now turn to the development of some technical properties of homomorphisms.

Proposition 1

Let $f:V \to W$ be a homomorphism. Then:

(a) $(O)f = O$;
(b) $(-x)f = -(x)f$;
(c) $(x_1 + \cdots + x_n)f = (x_1)f + \cdots + (x_n)f$.

Proof

(a) $(O)f = (O + O)f = (O)f + (O)f \Rightarrow O = (O)f$.
(b) $(-x)f = (-1 \cdot x)f = -1 \cdot (x)f = -(x)f$.
(c) The proof is by induction. We know the result is true for $n = 2$. Suppose it is true for $n - 1$, $n \geq 3$. Then

$$\begin{aligned}(x_1 + \cdots + x_n)f &= ((x_1 + \cdots x_{n-1}) + x_n)f \\ &= (x_1 + \cdots + x_{n-1})f + (x_n)f \\ &= (x_1)f + \cdots + (x_{n-1})f + (x_n)f.\end{aligned}$$

DEFINITION 2 Let $f: V \to W$ be a homomorphism. Then the *kernel of f*, denoted Ker (f), is defined by Ker $(f) = \{x \in V : (x)f = O\}$. The *image of f*, denoted Im (f), is defined by Im $(f) = \{y \in W : y = (x)f$ for some $x \in V\}$.

For instance, in the above examples of homomorphisms f, g, h, and d, we see that

Ker $(f) = \{(0, 0)\}$, Im $(f) = \{(\alpha, \beta, \gamma) \in \mathbb{R}^3 : \gamma = 0\}$;
Ker $(g) = \{(\alpha, \beta, \gamma) \in \mathbb{R}^3 : \alpha = \beta = 0\}$, Im $(g) = \{(\alpha, \beta, \gamma) \in \mathbb{R}^3 : \gamma = 0\}$;
Ker $(h) = \{(0, 0, 0)\}$, Im $(h) = \mathbb{R}^3$;
Ker $(d) = \{F \in D([0, 1]) : F(x) = \alpha \in \mathbb{R}, \forall x \in [0, 1]\}$.

Two additional examples are

(4) the *identity map* $1:V \to V$ for any vector space V given by $(x)1 = x$, $\forall x \in V$, and
(5) the *zero map* $0:V \to W$ for any V and W given by $(x)0 = O$.

One observes that in each case, the kernel as well as the image are subspaces. This is no mere coincidence, since we have:

Proposition 2

Let $f:V \to W$ be a homomorphism. Then:

(a) Ker $(f) < V$;
(b) Im $(f) < W$.

Proof

(a) $O \in \operatorname{Ker}(f)$ so $\operatorname{Ker}(f) \neq \varnothing$. If $x, y \in \operatorname{Ker}(f)$, $\alpha, \beta \in D$, then

$$\begin{aligned}(\alpha x + \beta y)f &= (\alpha x)f + (\beta y)f \\ &= \alpha(x)f + \beta(y)f \\ &= \alpha \cdot O + \beta \cdot O = O,\end{aligned}$$

and so $\alpha x + \beta y \in \operatorname{Ker}(f)$ and $\operatorname{Ker}(f) < V$.

(b) $O \in \operatorname{Im}(f)$ so $\operatorname{Im}(f) \neq \varnothing$. If $x, y \in \operatorname{Im}(f)$ there exist $u, v \in V$ such that $(u)f = x$, $(v)f = y$. If $\alpha, \beta \in D$ then

$$\begin{aligned}(\alpha u + \beta v)f = (\alpha u)f + (\beta v)f &= \alpha(u)f + \beta(v)f \\ &= \alpha x + \beta y \in \operatorname{Im}(f),\end{aligned}$$

and so $\operatorname{Im}(f) < W$.

DEFINITION 3 Let $f: V \to W$ be a homomorphism.

(a) If $\operatorname{Ker}(f) = O$, then f is called a *monomorphism* (f is *monic*).
(b) If $\operatorname{Im}(f) = W$, then f is called an *epimorphism* (f is *epic*).
(c) If f is both a monomorphism and an epimorphism, then f is called an *isomorphism*.

If there exists an isomorphism from V to W we say V is *isomorphic* to W, and write $V \approx W$. (See Exercise 5.)

Proposition 3

Let $f: V \to W$, $g: W \to U$ be homomorphisms.

(a) $fg: V \to U$ is a homomorphism.
(b) If f, g are monomorphisms, then fg is a monomorphism.
(c) If f, g are epimorphisms, then fg is an epimorphism.
(d) If f, g are isomorphisms, then fg is an isomorphism.

Proof

(a) If $x, y \in V$, then

$$\begin{aligned}(x + y)(fg) &= ((x + y)f)g \\ &= ((x)f + (y)f)g \\ &= ((x)f)g + ((y)f)g \\ &= (x)(fg) + (y)(fg).\end{aligned}$$

If $\alpha \in D$, then

$$\begin{aligned}(\alpha x)(fg) &= ((\alpha x)f)g \\ &= (\alpha(x)f)g \\ &= \alpha((x)f)g \\ &= \alpha(x)(fg).\end{aligned}$$

(b) $O = (x)(fg) = ((x)f)g \Rightarrow (x)f = O$, since g is a monomorphism. But $(x)f = O \Rightarrow x = O$, since f is a monomorphism.
(c) $u \in U \Rightarrow \exists w \in W$ such that $(w)g = u$. But $w \in W \Rightarrow \exists v \in V$ such that $(v)f = w$, and hence $(v)(fg) = u$.
(d) The proof follows immediately from (b) and (c).

EXERCISES

1. Determine which of the following maps of $\mathbb{R}^2$ into $\mathbb{R}^2$ are homomorphisms and find their images and kernels.

   (a) $(\alpha, \beta)f = (\alpha, 0)$.
   (b) $(\alpha, \beta)f = (\alpha\beta, \beta)$.
   (c) $(\alpha, \beta)f = (\alpha + \beta, \beta)$.
   (d) $(\alpha, \beta)f = (\alpha^2, \alpha)$.
   (e) $(\alpha, \beta)f = (1 + \alpha, \beta)$.
   (f) $(\alpha, \beta)f = (\beta, \alpha)$.

2. In the Introduction we called a function $f: S \to T$ *one-to-one*, denoted 1-1, if $(s)f = (s')f \Rightarrow s = s'$. Suppose f is a homomorphism from ${}_DV$ into ${}_DW$. Prove that f is 1-1 $\Leftrightarrow f$ is a monomorphism.

3. Recall the definition of the *inverse function* given in the Introduction. Let $f: V \to W$ be an isomorphism. Show that the inverse function $f^{-1}: W \to V$ is also an isomorphism. (You must show in particular that f^{-1} is a homomorphism.)

DEFINITION 4 A *group* consists of a nonempty set G together with a binary operation $G \times G \to G$ such that, denoting the image of the ordered pair (g_1, g_2) by $g_1 \cdot g_2$, the following conditions are satisfied:

(a) $g_1 \cdot (g_2 \cdot g_3) = (g_1 \cdot g_2) \cdot g_3$, $\forall g_1, g_2, g_3 \in G$;
(b) there exists $e \in G$ such that $g \cdot e = e \cdot g = g$, $\forall g \in G$;
(c) for each $g \in G$, $\exists g^{-1} \in G$ such that $g \cdot g^{-1} = g^{-1} \cdot g = e$.

We call e the *identity element* of G, and g^{-1} the *inverse* of g. If $g_1 \cdot g_2 = g_2 \cdot g_1$, $\forall g_1, g_2 \in G$, we say G is *commutative* (or *Abelian*).

4. Let $_DV$ be any vector space. Show that V is a commutative group with the binary operation "+."

5. Let $_DV$ be a vector space. Show that the set of all isomorphisms from V onto V is a group using composition of maps as the binary operation.
This group is sometimes called the *general linear group* on V, and denoted $GL(V)$. It is also referred to as Aut (V), since an isomorphism from a vector space onto *itself* is called an *automorphism*.

*6. Let $V = {}_{\mathbb{R}}\mathbb{R}^2$. For each $\gamma \in \mathbb{R}$, define a map $f_\gamma : V \to V$ by $((\alpha, \beta))f_\gamma = (\alpha + \gamma\beta, \beta)$.

(a) Show that, for each γ, $f_\gamma \in GL(V)$.
(b) Show that $f_\gamma \cdot f_{-\gamma} = f_{-\gamma} \cdot f_\gamma = I$, the identity map on V.
(c) Show that the set $S = \{f_\gamma : \gamma \in \mathbb{R}\}$ is a commutative group with composition as the binary operation. We say S is a group *under composition*.
(d) Describe geometrically the effect of f_1 and f_{-2} on the coordinate plane.

Remark: The maps $\{f_\gamma\}$ are examples of *shear transformations* of $\mathbb{R}^2$. We call S a *subgroup* of $GL(V)$ since (recall the definition of "subspace") S is a subset of $GL(V)$ and is a group using the *same* binary operation as $GL(V)$. [1.11]

7. Let D be any division ring. Show directly from the definitions that D is an Abelian group under addition, and D^1 is a group under multiplication, where D^1 is the set of nonzero elements of D. Note that D^1 is *not* a subgroup of D.

8. Let $_DV$ be a vector space, $H < V$, $f \in GL(V)$. H is said to be *invariant under* f if $x \in H \Rightarrow (x)f \in H$. If $I(H) = \{f \in GL(V) : H$ is invariant under $f\}$ show that $I(H)$ is a subgroup of $GL(V)$.

9. (a) Let $V = M_1 \oplus M_2$. Define $f : V \to M_1$ by $(x_1 + x_2)f = x_1$, $\forall x_1 + x_2 \in V$, where $x_1 \in M_1$, $x_2 \in M_2$. Show that f is an epimorphism of V onto M_1 with kernel M_2. We call f the *projection of V onto M_1 parallel to M_2*.
(b) Let $\mathbb{R}^2 = L_1 \oplus L_2$, where L_1 and L_2 are lines through O. Describe geometrically the projection map of $\mathbb{R}^2$ onto L_1 parallel to L_2.

SECTION 1.6
FUNDAMENTAL THEOREMS OF HOMOMORPHISMS

The word homomorphism, arising from the Greek ὁμός, meaning "same," and μορφή, meaning "form" or "shape," suggests that the structure of a vector space is preserved under the action of such a map. In this section we show how the structure of V is related to the structure of the image, Im (f), of V under a homomorphism f from V into some vector space.

DEFINITION 1 Let ${}_DV$ be a vector space and $H < V$. The set $x + H = \{x + h : h \in H\}$, for any $x \in V$, is called a *coset*, or an *affine subspace* of V. The element x is called a *representative* of the coset $x + H$.

For example, in any vector space V,

(a) If $H = O$, then each vector $x = x + O$ is a coset;
(b) if $H = V$, then $V = x + V$ is a coset, where x is arbitrary;
(c) if $x \in H$, for $H < V$, then $H = x + H = O + H$ is a coset.

An interesting fact is that the collection of all cosets $x + H$, where H is fixed and x ranges throughout V, can be made into a vector space by suitably defining scalar multiplication and vector addition. The following special cases of this construction are offered to illustrate geometrically what is going on.

In $\mathbb{R}^2$, let $H = \{(0, \beta) : \beta \in \mathbb{R}\}$, that is, the y-axis of the coordinate plane. For each $(\alpha_1, \alpha_2) \in \mathbb{R}^2$,

$$\begin{aligned}(\alpha_1, \alpha_2) + H &= \{(\alpha_1, \alpha_2) + (0, \beta) : \beta \in \mathbb{R}\} = \{(\alpha_1, 0) + (0, \alpha_2 + \beta) : \beta \in \mathbb{R}\} \\ &= \{(\alpha_1, 0) + (0, \beta') : \beta' \in \mathbb{R}\} = (\alpha_1, 0) + H.\end{aligned}$$

In other words, each such coset has a representative on the x-axis. As Figure 1.2 indicates, the collection $\overline{V}$ of all such cosets is precisely the collection of all vertical lines in the coordinate plane. We define addition and scalar multiplication on $\overline{V}$ by

$$((\alpha_1, 0) + H) + ((\alpha_2, 0) + H) = (\alpha_1 + \alpha_2, 0) + H,$$

and

$$\alpha((\alpha_1, 0) + H) = (\alpha\alpha_1, 0) + H, \forall \alpha, \alpha_1, \alpha_2 \in \mathbb{R}.$$

It is easy to check that $\overline{V}$ is then a vector space with $(0, 0) + H$ as the 0-vector, and $(-\alpha_1, 0) + H$ as the additive inverse of $(\alpha_1, 0) + H$. Moreover,

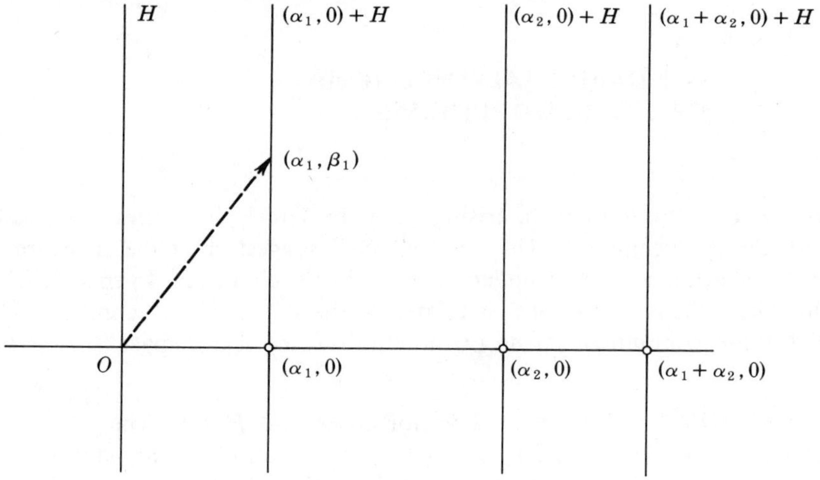

Figure 1.2

the obvious 1-1 function $(\alpha, 0) \to (\alpha, 0) + H$ is readily seen to be an isomorphism from the x-axis onto $\bar{V}$. Note also that the map $(\alpha, \beta) \to (\alpha, 0) + H$ is an epimorphism from V onto $\bar{V}$.

Likewise, in $V = \mathbb{R}^3$, if $H = \{(0, 0, \gamma) : \gamma \in \mathbb{R}\}$, that is, the z-axis, then the collection $\bar{V}$ of all cosets $x + H$ can be made into a vector space that turns out to be isomorphic to the "xy-plane." We begin by observing that $(\alpha_1, \beta_1, \gamma_1) + H = (\alpha_1, \beta_1, 0) + H$, as before, so that each coset has a representative in the xy-plane. As Figure 1.3 indicates, $\bar{V}$ is the totality of lines in space parallel to the z-axis. We define

$$((\alpha_1, \beta_1, 0) + H) + ((\alpha_2, \beta_2, 0) + H) = (\alpha_1 + \alpha_2, \beta_1 + \beta_2, 0) + H,$$
$$\alpha((\alpha_1, \beta_1, 0) + H) = (\alpha\alpha_1, \alpha\beta_1, 0) + H,$$

and the assertions follow by simple computations, with the isomorphism being $(\alpha, \beta, 0) \to (\alpha, \beta, 0) + H$. Again we assert that the map $(\alpha, \beta, \gamma) \to (\alpha, \beta, 0) + H$ is an epimorphism from V onto $\bar{V}$.

In each of the above examples, the arithmetic on the cosets appears to depend on the availability of suitable representatives. We shall see shortly that this is not the case. In terms of the first example, we could have used (α_1, β_1) as a representative of the coset $(\alpha_1, 0) + H = (\alpha_1, \beta_1) + H$, and used (α_2, β_2) as a representative of the coset $(\alpha_2, 0) + H = (\alpha_2, \beta_2) + H$, to arrive at

$$((\alpha_1, \beta_1) + H) + ((\alpha_2, \beta_2) + H) = (\alpha_1 + \alpha_2, \beta_1 + \beta_2) + H,$$

since the end result is the same: $(\alpha_1 + \alpha_2, \beta_1 + \beta_2) + H = (\alpha_1 + \alpha_2, 0) + H$, for any $\beta_1, \beta_2 \in \mathbb{R}$.

With this much intuitive introduction, let us develop the theory rigorously.

Proposition 1

Let ${}_DV$ be a vector space and $H < V$. For $x, y \in V$, the following statements are equivalent:

(a) $(x + H) \cap (y + H) \neq \varnothing$;
(b) $x - y \in H$;
(c) $x + H = y + H$.

Proof

(a) ⇒ (b) If $z \in (x + H) \cap (y + H)$, then there exist $h_1, h_2 \in H$ such that $z = x + h_1 = y + h_2 \Rightarrow x - y = h_2 - h_1 \in H$.

(b) ⇒ (c) Suppose $x - y = z \in H$. For $h \in H$, $x + h = (y + z) + h = y + (z + h) \in y + H$, since $z + h \in H$ and $x + H \subseteq y + H$. Conversely, $y + h = (x - z) + h = x + (h - z) \in x + H$ and $y + H \subseteq x + H$.

(c) ⇒ (a) The proof is obvious.

Corollary

If $y \in x + H$, then $y + H = x + H$.

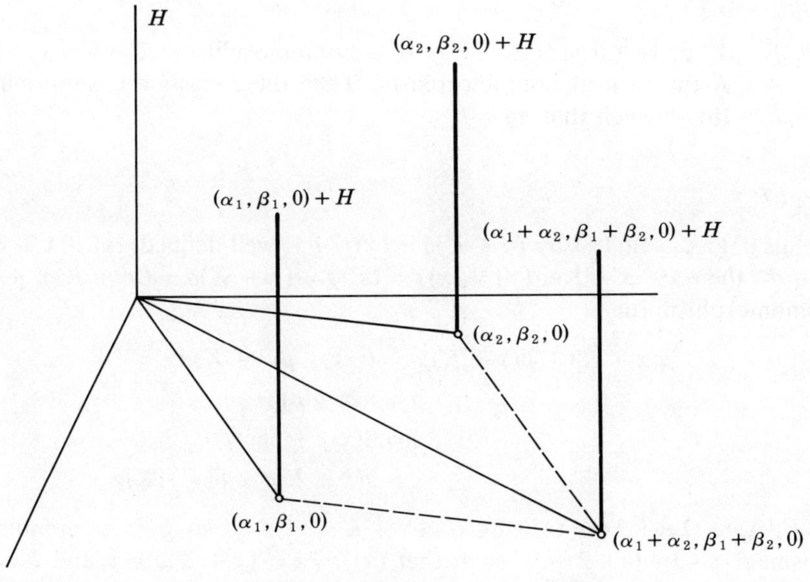

Figure 1.3

Proposition 2

Let ${}_DV$ be a vector space, $H < V$. Let $V/H = \{x + H : x \in V\}$. Under the definitions $(x + H) + (y + H) = (x + y) + H$ and $\alpha(x + H) = (\alpha x + H)$, V/H is a vector space.

Proof

We must first show that the addition and scalar multiplication of cosets is well defined. Suppose $x + H = x' + H$ and $y + H = y' + H$. We must show $(x + y) + H = (x' + y') + H$ and $\alpha x + H = \alpha x' + H$. $x + H = x' + H \Rightarrow x - x' = h_1 \in H$ and $y + H = y' + H \Rightarrow y - y' = h_2 \in H$. Therefore $(x + y) - (x' + y') = (x - x') + (y - y') = h_1 + h_2 \in H \Rightarrow (x + y) + H = (x' + y') + H$ and $\alpha(x - x') = \alpha h_1 \in H \Rightarrow \alpha x + H = \alpha x' + H$.

With the comment that the additive identity of V/H is $O + H$ and the inverse of $x + H$ is $-x + H$, the remainder of the proof is left to the reader.

DEFINITION 2 Let ${}_DV$ be a vector space, $H < V$; then $\pi : V \to V/H$ defined by $(x)\pi = x + H$ is called the *natural homomorphism of V onto V/H.*

The reader should note that our calling π a homomorphism is not mere optimism; it is a consequence of the definition of addition and scalar multiplication in V/H that shows that π is a homomorphism.

Theorem 1

Let ${}_DV$, ${}_DW$ be vector spaces, $f : V \to W$ a homomorphism, $K = \text{Ker}(f)$ and $\pi : V \to V/K$ the natural homomorphism. Then there exists an isomorphism $g : V/K \to \text{Im}(f)$ such that $\pi g = f$.

Proof

Define $g : V/K \to \text{Im}(f)$ by $(x + K)g = (x)f$. g is well-defined, for if $x + K = x' + K$, then $x - x' \in \text{Ker}(f) \Rightarrow (x)f = (x')f \Rightarrow (x + K)g = (x' + K)g$. g is a homomorphism for

$$\begin{aligned}(\alpha(x + K) + \beta(y + K))g &= ((\alpha x + \beta y) + K)g \\ &= (\alpha x + \beta y)f \\ &= \alpha(x)f + \beta(y)f \\ &= \alpha(x + K)g + \beta(y + K)g.\end{aligned}$$

$(x + K)g = O \Rightarrow (x)f = O \Rightarrow x \in K \Rightarrow x + K = O + K$, so g is a monomorphism. If $y \in \text{Im}(f)$, $\exists x \in V$ such that $(x)f = y \Rightarrow (x + K)g = y$ and g is an isomorphism.

As an interesting example of this theorem in action, recall Example (3) of Section 1.5 and let $d:D([0, 1]) \to F([0, 1])$ be the differentiation map. Ker (d) are the constant functions that are identified with $\mathbb{R}$, and the fundamental theorem of homomorphisms says there exists an isomorphism $V/\mathbb{R} \to \text{Im}\,(d)$. Being an isomorphism it has an inverse, and this inverse is called *antidifferentiation* or *indefinite integration*.

We conclude this section with the promised result relating the structure of two vector spaces one of which is a homomorphic image of the other. Recall that if $f:S \to T$ is a function from a set S to a set T, then for $S' \subseteq S$ we define $(S')f = \{(s)f:s \in S'\}$.

Theorem 2

Let $f:V \to W$ be an epimorphism with kernel K. Let $\mathscr{L} = \{U \subseteq V:K \subseteq U < V\}$, and $\mathscr{L}' = \{U' \subseteq W : U' < W\}$. Then:

(a) $U \in \mathscr{L} \Rightarrow (U)f \in \mathscr{L}'$;
(b) $U' \in \mathscr{L}' \Rightarrow U' = (U)f$ for some unique $U \in \mathscr{L}$;
(c) $U_1, U_2 \in \mathscr{L}$, and $U_1 \subseteq U_2 \Rightarrow (U_1)f \subseteq (U_2)f$;
(d) $U_1, U_2 \in \mathscr{L} \Rightarrow U_1 \cap U_2 \in \mathscr{L}$, and $(U_1 \cap U_2)f = (U_1)f \cap (U_2)f$;
(e) $U_1, U_2 \in \mathscr{L} \Rightarrow U_1 + U_2 \in \mathscr{L}$, and $(U_1 + U_2)f = (U_1)f + (U_2)f$.

In short, there is a 1-1 correspondence between the subspaces of V that contain K and the subspaces of W; this correspondence preserves inclusion, sum, and intersection.

Proof

(a) Let $U \in \mathscr{L}$. Let $(u_1)f, (u_2)f \in (U)f$, where $u_1, u_2 \in U$. For any $\alpha, \beta \in D$, we have $\alpha(u_1)f + \beta(u_2)f = (\alpha u_1 + \beta u_2)f \in (U)f$, since $\alpha u_1 + \beta u_2 \in U$.
(b) Let $U' \in \mathscr{L}'$. Put $U = \{u \in V:(u)f \in U'\}$. Then $U \neq \varnothing$, since $O \in U$. If $u_1, u_2 \in U$ and $\alpha, \beta \in D$, then $(\alpha u_1 + \beta u_2)f = \alpha(u_1)f + \beta(u_2)f \in U'$, since U' is a subspace. Hence $\alpha u_1 + \beta u_2 \in U$, and $U < V$. Also, $K \subseteq U$ since $(x)f = O \in U'$, for any $x \in K$. The uniqueness of U can be shown as follows: suppose $K < U_1 < V$, and $(U_1)f = U'$. Then $U_1 \subseteq U$, by the definition of U. Also if $u \in U$, then $(u)f = (u_1)f$, for some $u_1 \in U_1$. Hence $(u - u_1)f = O$ so that $u - u_1 = k \in K$. Then $u = u_1 + k \in U_1 + K = U_1$, showing $U \subseteq U_1$.
(c) The proof is obvious.
(d) Let $U_1, U_2 \in \mathscr{L}$. We have shown earlier that $U_1 \cap U_2 < V$. It is also clear from set theory that $K \subseteq U_1 \cap U_2$, and $(U_1 \cap U_2)f \subseteq (U_1)f \cap (U_2)f$. Now let $(u_1)f = (u_2)f \in (U_1)f \cap (U_2)f$. Then $(u_1 - u_2)f = O$, so that $u_1 - u_2 = x \in K$. Hence $u_1 = u_2 + x \in U_1 \cap U_2$, and so $(u_1)f \in (U_1 \cap U_2)f$, giving the reverse inclusion.

(e) Let $U_1, U_2 \in \mathscr{L}$. Again we know that $U_1 + U_2 < V$, and that $K \subseteq U_1 + U_2$. Finally,

$$\begin{aligned}(U_1 + U_2)f &= \{(u_1 + u_2)f : u_1 \in U_1, u_2 \in U_2\} \\ &= \{(u_1)f + (u_2)f : u_1 \in U_1, u_2 \in U_2\} \\ &= (U_1)f + (U_2)f,\end{aligned}$$

and the theorem is proved.

Corollary

Let $H < V$. The function given by $U \to \{u + H : u \in U\}$ is a 1-1 correspondence between the set of subspaces of V that contain H, and the set of subspaces of V/H.

Proof

If π is the natural map, then $(U)\pi = \{u + H : u \in U\}$.

Theorem 1 is frequently called the *Fundamental Theorem of Homomorphisms*. Sometimes Theorem 1 as well as variations and extensions of Theorem 2 are collectively referred to by this name.

EXERCISES

1. Let $(\alpha_1, \beta_1), (\alpha_2, \beta_2) \in \mathbb{R}^2$ be distinct. Show that $\{\lambda(\alpha_1, \beta_1) + (1 - \lambda)(\alpha_2, \beta_2) : \lambda \in \mathbb{R}\}$ is a coset relative to the subspace $\langle(\alpha_1 - \alpha_2, \beta_1 - \beta_2)\rangle$.

2. Define $f : \mathbb{R}^3 \to \mathbb{R}^3$ by $(\alpha, \beta, \gamma)f = (\alpha, \gamma, 0)$. Find Ker (f) and describe the coset of $(3, 1, -1)$.

3. Let $L = \langle(\alpha_1, \beta_1)\rangle$ be any "line" in $\mathbb{R}^2$ other than $\langle(0, 1)\rangle = H$, so that $L \oplus H = \mathbb{R}^2$.

   (a) Show that each coset of the form $x + H$ intersects L in precisely one element.

   (b) Sketch a picture analogous to Figure 1.2 with L playing the role of the x-axis, and indicate how addition of cosets is performed using representatives from L.

4. Let $L \oplus H = V$, for some vector space V and subspaces L, H. Show that:

   (a) for $x \in V$, $x + H = l + H$ where $x = l + h$, $l \in L$, $h \in H$;

   (b) using the notation of (a), $(x + H) \cap L = l$;

(c) $(x + y + H) \cap L = ((x + H) \cap L) + ((y + H) \cap L)$;
(d) $(\alpha x + H) \cap L = \alpha \cdot ((x + H) \cap L)$;
(e) $V/H \approx L$.

5. Complete the proof of Proposition 2.

6. Show that the natural homomorphism is a homomorphism (Definition 2).

SECTION 1.7
THE RELATIONSHIP BETWEEN HOMOMORPHISMS AND BASES

We turn now to the relationship between homomorphisms from V into W and bases of V. Specifically, we shall prove the existence and uniqueness of a homomorphism $f: V \to W$ that has prescribed values $(v_1)f, \cdots, (v_n)f$ in W for some basis $\{v_1, \cdots, v_n\}$ of V.

The reader may recall an analogous situation in analytic geometry. There is a unique line passing through two given points in the plane. There is a unique parabola (perhaps a line) with vertical axis through three points in the plane. In terms of the polynomial functions whose graphs are lines, parabolas, and so on, there is a unique polynomial p of degree less than or equal to n that has prescribed values $p(\alpha_1), \cdots, p(\alpha_{n+1})$ for distinct real numbers $\alpha_1, \cdots, \alpha_{n+1}$. The formula which establishes this general assertion is called the *Lagrange Interpolation Formula*.

Of necessity we will be dealing with finite dimensional vector spaces throughout this section. The main results, however, carry over to infinite dimensional vector spaces, once the notions of basis and dimension have been properly introduced.

Proposition 1

Let ${}_DV$, ${}_DW$ be vector spaces, $B = \{x_1, \cdots, x_n\}$ a basis for V and $f, g: V \to W$ homomorphisms. Then $f = g \Leftrightarrow (x_i)f = (x_i)g$ for $i = 1, \cdots, n$.

Proof

$\boxed{\Rightarrow}$: This is true by the definition of equality of maps.

$\boxed{\Leftarrow}$: If $x \in V$, then there exist $\alpha_1, \cdots, \alpha_n \in D$ such that $x = \alpha_1 x_1 + \cdots + \alpha_n x_n$; then

$$\begin{aligned}(x)f &= (\alpha_1 x_1 + \cdots + \alpha_n x_n)f \\ &= (\alpha_1 x_1)f + \cdots + (\alpha_n x_n)f \\ &= \alpha_1(x_1)f + \cdots + \alpha_n(x_n)f \\ &= \alpha_1(x_1)g + \cdots + \alpha_n(x_n)g \\ &= (\alpha_1 x_1)g + \cdots + (\alpha_n x_n)g \\ &= (\alpha_1 x_1 + \cdots + \alpha_n x_n)g \\ &= (x)g\end{aligned}$$

and

$$f = g.$$

Theorem 1

Let ${}_DV$, ${}_DW$ be vector spaces, $B = \{x_1, \cdots, x_n\}$ a basis of V, $\{y_1, \cdots, y_n\} \subseteq W$. Then there exists a unique homomorphism $f: V \to W$ such that $(x_i)f = y_i$, $i = 1, \cdots, n$.

Proof

If $x \in V$ there exist unique $\alpha_1, \cdots, \alpha_n \in D$ such that $x = \alpha_1 x_1 + \cdots + \alpha_n x_n$. Define $f: V \to W$ by $(x)f = \alpha_1 y_1 + \cdots + \alpha_n y_n$. The uniqueness of the basis representation assures us that f is well defined and clearly $(x_i)f = y_i$ for $i = 1, \cdots, n$.

Proposition 1 tells us that there is at most one homomorphism with this property, so if we show that f is a homomorphism we are done.

If $\alpha \in D$ then $\alpha x = \alpha\alpha_1 x_1 + \cdots + \alpha\alpha_n x_n$ so

$$\begin{aligned}(\alpha x)f &= \alpha\alpha_1 y_1 + \cdots + \alpha\alpha_n y_n \\ &= \alpha(\alpha_1 y_1 + \cdots + \alpha_n y_n) = \alpha(x)f.\end{aligned}$$

If $y = \beta_1 x_1 + \cdots + \beta_n x_n \in V$ then $x + y = (\alpha_1 + \beta_1)x_1 + \cdots + (\alpha_n + \beta_n)x_n$, so

$$\begin{aligned}(x+y)f &= (\alpha_1 + \beta_1)y_1 + \cdots + (\alpha_n + \beta_n)y_n \\ &= (\alpha_1 y_1 + \cdots + \alpha_n y_n) + (\beta_1 y_1 + \cdots + \beta_n y_n) \\ &= (x)f + (y)f.\end{aligned}$$

Theorem 1 says that we can describe a homomorphism merely by prescribing the image of a basis, a technique we shall generally use later in this book.

Theorem 2

Let ${}_DV$, ${}_DW$ be finite dimensional vector spaces. Then $\dim(V) = \dim(W) \Leftrightarrow V \approx W$.

Proof

$\boxed{\Rightarrow}$: Dim $(V) = \dim(W)$ implies that a basis of V has the same number of elements as a basis for W. Let $\{x_1, \cdots, x_n\}$ be a basis for V, $\{y_1, \cdots, y_n\}$ a basis for W. By Theorem 1 there exists a unique homomorphism $f: V \to W$ such that $(x_i)f = y_i$, $i = 1, \cdots, n$. If $x \in V$, $x = \alpha_1 x_1 + \cdots + \alpha_n x_n$ for suitable $\alpha_1, \cdots, \alpha_n \in D$. If $x \in \text{Ker}(f)$, then $O = (x)f = \alpha_1 y_1 + \cdots + \alpha_n y_n \Rightarrow \alpha_1 = \cdots = \alpha_n = 0$ since $y_1, \cdots, y_n$ are linearly independent. Therefore Ker $(f) = O$. If $y \in W$, then there exist $\beta_1, \cdots, \beta_n \in D$ such that

$$
\begin{aligned}
y = \beta_1 y_1 + \cdots + \beta_n y_n &= \beta_1(x_1)f + \cdots + \beta_n(x_n)f \\
&= (\beta_1 x_1 + \cdots + \beta_n x_n)f \in \text{Im}(f)
\end{aligned}
$$

and f is an isomorphism.

$\boxed{\Leftarrow}$: Let $f: V \to W$ be an isomorphism and $\{x_1, \cdots, x_n\}$ a basis of V. We claim that $\{(x_1)f, \cdots, (x_n)f\}$ is a basis for W. If $y \in W$ there exists $x \in V$ such that $(x)f = y$. For some $\alpha_1, \cdots, \alpha_n \in D$, $x = \alpha_1 x_1 + \cdots + \alpha_n x_n$ and hence $y = (x)f = \alpha_1(x_1)f + \cdots + \alpha_n(x_n)f$, which implies that $\{(x_1)f, \cdots, (x_n)f\}$ spans W. To show linear independence, suppose we have $\beta_1, \cdots, \beta_n \in D$ such that $O = \beta_1(x_1)f + \cdots + \beta_n(x_n)f$. Then $O = (\beta_1 x_1 + \cdots + \beta_n x_n)f$, which, since f is a monomorphism, implies $\beta_1 x_1 + \cdots + \beta_n x_n = O \Rightarrow \beta_1 = \cdots = \beta_n = 0$.

We record as corollaries several elementary, but often useful, facts that were demonstrated in the proof of Theorem 2.

Corollary 1

Let ${}_DV$, ${}_DW$ be finite dimensional vector spaces, and $f: V \to W$ a homomorphism. f is an isomorphism $\Leftrightarrow$ the image of a basis of V is a basis of W.

Corollary 2

If $f: V \to W$ is a monomorphism, then the image of a linearly independent set is linearly independent.

Corollary 3

If $f: V \to W$ is an epimorphism, then the image of a generating set is a generating set.

Theorem 3

Let ${}_DV$, ${}_DW$ be finite dimensional vector spaces and $f:V \to W$ a homomorphism. Then $\dim(V) = \dim(\text{Ker}(f)) + \dim(\text{Im}(f))$.

Proof

Let $\{x_1, \cdots, x_m\}$ be a basis of Ker (f) and extend it to a basis $\{x_1, \cdots, x_m, y_1, \cdots, y_{n-m}\}$ of V. We claim that $\{(y_1)f, \cdots, (y_{n-m})f\}$ is a basis of Im (f). If $\alpha_1(y_1)f + \cdots + \alpha_{n-m}(y_{n-m})f = O$, then

$$\alpha_1 y_1 + \cdots + \alpha_{n-m} y_{n-m} \in \text{Ker}(f) = \langle x_1, \cdots, x_m \rangle \Rightarrow \alpha_1 = \cdots = \alpha_{n-m} = 0,$$

and so $\{(y_1)f, \cdots, (y_{n-m})f\}$ is linearly independent. If $w \in \text{Im}(f)$ there exists $z \in V$ with $(z)f = w$. But $z = \alpha_1 x_1 + \cdots + \alpha_m x_m + \beta_1 y_1 + \cdots + \beta_{n-m} y_{n-m}$ so

$$\begin{aligned} w = (z)f &= \alpha_1(x_1)f + \cdots + \alpha_m(x_m)f + \beta_1(y_1)f + \cdots + \beta_{n-m}(y_{n-m})f \\ &= \beta_1(y_1)f + \cdots + \beta_{n-m}(y_{n-m})f, \end{aligned}$$

and we are done.

Corollary

Let $H < V$ with V finite dimensional. Then $\dim(V/H) = \dim(V) - \dim(H)$, and hence $V/H \approx H'$ whenever $V = H \oplus H'$.

EXERCISES

1. Let $e_1 = (1, 0, 0)$, $e_2 = (0, 1, 0)$, $e_3 = (0, 0, 1)$ in $\mathbb{R}^3$. Define $f:\mathbb{R}^3 \to \mathbb{R}^4$ by $(e_1)f = (1, 0, 0, 1)$, $(e_2)f = (0, 1, 0, 1)$, $(e_3)f = (1, 1, 0, 1)$.

   (a) Find $(-1, 0, 1)f$, $(2, 1, 1)f$, and $(3, 3, 1)f$.
   (b) Find Im (f) and Ker (f).

2. Give a geometric description of each of the following linear maps from $\mathbb{R}^2$ into $\mathbb{R}^2$:

   (a) $((1, 0))f = (0, 1)$, $((0, 1))f = (1, 0)$.
   (b) $((1, 0))g = (1, 0)$, $((0, 1))g = (0, -1)$.
   (c) $((1, 0))h = (2, 0)$, $((0, 1))h = (0, 3)$.
   (d) $((1, 1))k = (-1, 1)$, $((-1, 1))k = (1, 1)$.
   (e) $((1, 0))p = (\frac{1}{2}, 0)$, $((0)), 1p = (0, 0)$.

3. Let V be a finite dimensional vector space, and $H < V$, $H \neq V$. Show that if $x, y \in V$ and $x, y \notin H$, then $\exists f \in GL(V)$ such that $(x)f = y$, and $(h)f = h$, $\forall h \in H$. [*Hint*: Extend a basis of H in two ways to a basis of V.]

4. In the proof of Theorem 1 it was essential that $\{x_1, \cdots, x_n\}$ be a basis. To see this, let $V = W = \mathbb{R}^2$, $x_1 = (1, 0)$, $x_2 = (0, 1)$, $x_3 = (1, 1)$, $y_1 = (1, 2)$, $y_2 = (1, 0)$, $y_3 = (3, 2)$. Show that there exists no homomorphism $f:\mathbb{R}^2 \to \mathbb{R}^2$ such that $(x_i)f = y_i$ for $i = 1, 2, 3$.

5. The generalization of Theorem 1 hinted at in Exercise 4 is as follows. Let ${}_DV$, ${}_DW$ be vector spaces, $S = \{x_1, \cdots, x_n\}$ a spanning set for V, and $T = \{y_1, \cdots, y_n\} \subseteq W$. Recalling the notation of Exercise 6, Section 1.3, show that there exists a homomorphism $f:V \to W$ such that $(x_i)f = y_i$, $i = 1, \cdots, n \Leftrightarrow \mathscr{D}(S) \subseteq \mathscr{D}(T)$. Can you remove the condition that $\langle S \rangle = V$ by requiring that V be finite dimensional?

6. Let $V = \mathbb{R}^2$. For $0 \leq \theta < 2\pi$ the homomorphism $f_\theta:V \to V$ defined by $(1, 0)f_\theta = (\cos\theta, \sin\theta)$ and $(0,1)f_\theta = (-\sin\theta, \cos\theta)$ is a *rotation*. Show that f_θ is an isomorphism and give its inverse. Show that $\{f_\theta:0 \leq \theta < 2\pi\} = H$ is a subgroup of $GL(V)$.

*7. Let ${}_DV$ be a finite dimensional vector space and $f:V \to V$ a homomorphism. Show that f is a monomorphism $\Leftrightarrow f$ is an epimorphism. The fact that V is finite dimensional is crucial to this proof. Let $d:\mathbb{R}[x] \to \mathbb{R}[x]$ be the differentiation map and show that d is an epimorphism but not an isomorphism. Can you find a homomorphism of $\mathbb{R}[x]$ to itself that when followed by d yields the identity but when preceded by d does not? [1.9]

SECTION 1.8
MATRICES

It is the intent of the authors to approach linear algebra and geometry in as conceptual a manner as pedagogical realities permit. However, computational methods cannot be completely ignored, for computation is frequently a prelude to understanding concepts and an aid in their development. Matrices are one of the principal computational tools of linear algebra.

DEFINITION 1 An $n \times m$ *matrix* with entries from the division ring D is a rectangular array with n rows and m columns of elements of D, that is,

$$\begin{pmatrix} \alpha_{11} & \alpha_{12} & \cdots & \alpha_{1m} \\ \alpha_{21} & \alpha_{22} & \cdots & \alpha_{2m} \\ \vdots & \vdots & & \vdots \\ \alpha_{n1} & \alpha_{n2} & \cdots & \alpha_{nm} \end{pmatrix}.$$

The above matrix will frequently be denoted by (α_{ij}). The set of all $n \times m$ matrices with entries from D will be denoted $D_{n \times m}$. In particular $D_{n \times n}$ will be denoted D_n. If $(\alpha_{ij}), (\beta_{ij}) \in D_{n \times m}$ we define $(\alpha_{ij}) = (\beta_{ij})$ if $\alpha_{ij} = \beta_{ij}$ for all $i, j, 1 \leq i \leq n, 1 \leq j \leq m$.

There is a relation between matrices and vector-space homomorphisms that occurs as follows. Let ${}_DV$, ${}_DW$ be vector spaces, $B = \{x_1, \cdots, x_n\}$ a basis of V, $C = \{y_1, \cdots, y_m\}$ a basis of W, and $f: V \to W$ a homomorphism. If $(x_i)f = \alpha_{i1}y_1 + \cdots + \alpha_{im}y_m$ for $i = 1, \cdots, n$, we write

$$m(f, B, C) = \begin{pmatrix} \alpha_{11} & \alpha_{12} & \cdots & \alpha_{1m} \\ \alpha_{21} & \alpha_{22} & \cdots & \alpha_{2m} \\ \vdots & \vdots & & \vdots \\ \alpha_{n1} & \alpha_{n2} & \cdots & \alpha_{nm} \end{pmatrix}.$$

An important, though simple, example occurs when $V = W$, $B = C$ and we have the map $1: V \to V$ defined by $(x)1 = x$ for all $x \in V$. Then

$$m(1, B, B) = \begin{pmatrix} 1 & 0 & \cdots & 0 \\ 0 & 1 & \cdots & 0 \\ \vdots & & & \vdots \\ 0 & 0 & \cdots & 1 \end{pmatrix}.$$

This matrix is called the *identity matrix* of D_n and denoted I_n, or I where the context makes n clear. In a case such as this where $V = W$ and $B = C$, we shall write $m(f, B)$ for $m(f, B, B)$.

Conversely, given V, W, B, C as above and $(\beta_{ij}) \in D_{n \times m}$, Theorem 1, Section 1.7 tells us that there exists a unique homomorphism $g: V \to W$ such that $(x_i)g = \beta_{i1}y_1 + \cdots + \beta_{im}y_m$ and hence $m(g, B, C) = (\beta_{ij})$. If we denote by $\text{Hom}_D(V, W)$ the set of all homomorphisms from V to W, the foregoing discussion establishes:

Proposition 1

If ${}_DV$, ${}_DW$ are finite dimensional vector spaces, $B = \{x_1, \cdots, x_n\}$ a basis of V, $C = \{y_1, \cdots, y_m\}$ a basis of W, then the map $f \to m(f, B, C)$ is a 1-1 map of $\text{Hom}_D(V, W)$ onto $D_{n \times m}$.

A comment should be made about this proposition. In the definition of the function $f \to m(f, B, C)$, bases are used in a different way than they have been in the past. Prior to this time bases have been sets with certain properties, but now the *order* in which the elements are written makes a difference, since reordering the elements of B interchanges rows of the matrix and reordering elements of C interchanges columns of the matrix. Whenever matrices are being discussed and are being related to vector-space homomorphisms via a basis, we shall tacitly assume that the basis is ordered.

Having defined matrices and established their relation to homomorphisms, the next item on the agenda is to introduce matrix operations that correspond to operations on homomorphisms. We begin by defining the addition of homomorphisms.

Proposition 2

Let $f:V \to W$ and $g:V \to W$ be vector-space homomorphisms. The map $f + g:V \to W$ defined by $(x)(f + g) = (x)f + (x)g$ for all $x \in V$ is a vector-space homomorphism.

Proof

If $x, y \in V$, then

$$\begin{aligned}(x + y)(f + g) &= (x + y)f + (x + y)g \\ &= (x)f + (y)f + (x)g + (y)g \\ &= (x)f + (x)g + (y)f + (y)g \\ &= (x)(f + g) + (y)(f + g).\end{aligned}$$

If $\alpha \in D$, then

$$\begin{aligned}(\alpha x)(f + g) &= (\alpha x)f + (\alpha x)g \\ &= \alpha(x)f + \alpha(x)g \\ &= \alpha((x)f + (x)g) \\ &= \alpha(x)(f + g).\end{aligned}$$

Let ${}_DV$, ${}_DW$ be vector spaces, $B = \{x_1, \cdots, x_n\}$ a basis for V, $C = \{y_1, \cdots, y_m\}$ a basis for W, and $f, g:V \to W$ vector-space homomorphisms. We wish to define the addition of matrices in such a way that $m(f, B, C) + m(g, B, C) = m(f + g, B, C)$. If $(x_i)f = \alpha_{i1}y_1 + \cdots + \alpha_{im}y_m$ and $(x_i)g = \beta_{i1}y_1 + \cdots + \beta_{im}y_m$, then

$$\begin{aligned}(x_i)(f + g) &= (x_i)f + (x_i)g \\ &= \alpha_{i1}y_1 + \cdots + \alpha_{im}y_m + \beta_{i1}y_1 + \cdots + \beta_{im}y_m \\ &= (\alpha_{i1} + \beta_{i1})y_1 + \cdots + (\alpha_{im} + \beta_{im})y_m.\end{aligned}$$

Hence, if we are to have $m(f, B, C) + m(g, B, C) = m(f + g, B, C)$, our definition should be the following:

DEFINITION 2 If

$$\begin{pmatrix} \alpha_{11} & \alpha_{12} & \cdots & \alpha_{1m} \\ \alpha_{21} & \alpha_{22} & \cdots & \alpha_{2m} \\ \vdots & \vdots & & \vdots \\ \alpha_{n1} & \alpha_{n2} & \cdots & \alpha_{nm} \end{pmatrix}, \quad \begin{pmatrix} \beta_{11} & \beta_{12} & \cdots & \beta_{1m} \\ \beta_{21} & \beta_{22} & \cdots & \beta_{2m} \\ \vdots & \vdots & & \vdots \\ \beta_{n1} & \beta_{n2} & \cdots & \beta_{nm} \end{pmatrix} \in D_{n \times m},$$

then we define

$$\begin{pmatrix} \alpha_{11} & \alpha_{12} & \cdots & \alpha_{1m} \\ \alpha_{21} & \alpha_{22} & \cdots & \alpha_{2m} \\ \vdots & \vdots & & \vdots \\ \alpha_{n1} & \alpha_{n2} & \cdots & \alpha_{nm} \end{pmatrix} + \begin{pmatrix} \beta_{11} & \beta_{12} & \cdots & \beta_{1m} \\ \beta_{21} & \beta_{22} & \cdots & \beta_{2m} \\ \vdots & \vdots & & \vdots \\ \beta_{n1} & \beta_{n2} & \cdots & \beta_{nm} \end{pmatrix}$$

$$= \begin{pmatrix} \alpha_{11} + \beta_{11} & \alpha_{12} + \beta_{12} & \cdots & \alpha_{1m} + \beta_{1m} \\ \alpha_{21} + \beta_{21} & \alpha_{22} + \beta_{22} & \cdots & \alpha_{2m} + \beta_{2m} \\ \vdots & \vdots & & \vdots \\ \alpha_{n1} + \beta_{n1} & \alpha_{n2} + \beta_{n2} & \cdots & \alpha_{nm} + \beta_{nm} \end{pmatrix}$$

or, briefly, $(\alpha_{ij}) + (\beta_{ij}) = (\alpha_{ij} + \beta_{ij})$.

You should note that we have only defined the addition of two matrices of the same size, and in the definition of multiplication that follows you should consider the significance of the relative size of the factors.

Suppose ${}_DU$, ${}_DV$, ${}_DW$ are vector spaces, $A = \{z_1, \cdots, z_r\}$ a basis of U, $B = \{x_1, \cdots, x_n\}$ a basis of V, $C = \{y_1, \cdots, y_m\}$ a basis of W, $g: U \to V$, $f: V \to W$ homomorphisms. We wish to define the product of matrices so that $m(g, A, B) \cdot m(f, B, C) = m(gf, A, C)$.

Suppose $(z_i)g = \beta_{i1}x_1 + \cdots + \beta_{in}x_n$, $1 \le i \le r$, and

$$(x_i)f = \alpha_{i1}y_1 + \cdots + \alpha_{im}y_m, \quad 1 \le i \le n.$$

Then

$$\begin{aligned} (z_i)(gf) = ((z_i)g)f &= \left(\sum_{j=1}^{n} \beta_{ij} x_j\right) f \\ &= \sum_{j=1}^{n} \beta_{ij}(x_j)f = \sum_{j=1}^{n} \beta_{ij}\left(\sum_{k=1}^{m} \alpha_{jk} y_k\right) \\ &= \sum_{j=1}^{n} \sum_{k=1}^{m} \beta_{ij} \alpha_{jk} y_k = \sum_{k=1}^{m} \left(\sum_{j=1}^{n} \beta_{ij} \alpha_{jk}\right) x_k, \end{aligned}$$

which suggests the following:

DEFINITION 3 Let $(\beta_{ij}) \in D_{r\times n}$ and $(\alpha_{ij}) \in D_{n\times m}$. Then $(\beta_{ij})(\alpha_{ij}) = (\gamma_{ij}) \in D_{r\times m}$, where $\gamma_{ij} = \sum_{k=1}^{n} \beta_{ik}\alpha_{kj}$.

We illustrate the addition and multiplication of matrices with the following examples.

EXAMPLE 1

$$\begin{pmatrix} 1 & -1 & 2 \\ 3 & 0 & -1 \end{pmatrix} + \begin{pmatrix} 2 & 0 & -1 \\ 1 & 1 & 0 \end{pmatrix} = \begin{pmatrix} 1+2 & -1+0 & 2-1 \\ 3+1 & 0+1 & -1+0 \end{pmatrix}$$
$$= \begin{pmatrix} 3 & -1 & 1 \\ 4 & 1 & -1 \end{pmatrix}.$$

EXAMPLE 2

$$\begin{pmatrix} 1 & -1 & 2 \\ 3 & 0 & -1 \end{pmatrix} \cdot \begin{pmatrix} 2 & 1 \\ 0 & 1 \\ -1 & 0 \end{pmatrix} = \begin{pmatrix} 1\cdot 2 - 1\cdot 0 + 2\cdot(-1) & 1\cdot 1 - 1\cdot 1 + 2\cdot 0 \\ 3\cdot 2 + 0\cdot 0 - 1\cdot(-1) & 3\cdot 1 + 0\cdot 1 - 1\cdot 0 \end{pmatrix}$$
$$= \begin{pmatrix} 0 & 0 \\ 7 & 3 \end{pmatrix}.$$

As a further illustration of the definitions of multiplication of matrices and equality of matrices we prove:

Proposition 3

The multiplication of matrices is associative, that is, if $(\alpha_{ij}) \in D_{n\times m}$, $(\beta_{ij}) \in D_{m\times r}$, $(\gamma_{ij}) \in D_{r\times s}$, then

$$[(\alpha_{ij})(\beta_{ij})](\gamma_{ij}) = (\alpha_{ij})[(\beta_{ij})(\gamma_{ij})].$$

Proof

$$[(\alpha_{ij})(\beta_{ij})](\gamma_{ij}) = \left(\sum_{k=1}^{m} \alpha_{ik}\beta_{kj}\right)(\gamma_{ij})$$
$$= \left(\sum_{t=1}^{r}\sum_{k=1}^{m} [\alpha_{ik}\beta_{kt}]\gamma_{tj}\right)$$
$$= \left(\sum_{k=1}^{m}\sum_{t=1}^{r} \alpha_{ik}[\beta_{kt}\gamma_{tj}]\right)$$
$$= (\alpha_{ij})\left(\sum_{t=1}^{r} \beta_{it}\gamma_{tj}\right)$$
$$= (\alpha_{ij})[(\beta_{ij})(\gamma_{ij})].$$

This proof nicely illustrates the use of the definition of equality of matrices and the way in which some of the arithmetic properties of D are inherited by matrices. It is of course quite unnecessary, since the result follows immediately from the associativity of the composition of maps.

Proposition 4

Let D be a division ring, and identify D with D_1, the set of 1×1 matrices with entries from D.

(a) ${}_D D_{1 \times n}$ is a vector space where addition is matrix addition and scalar multiplication is the product of a 1×1 matrix with a $1 \times n$ matrix.
(b) Let ${}_D V$ be a vector space and $B = \{x_1, \cdots, x_n\}$ a basis for V. If $x \in V$ then $x = \alpha_1 x_1 + \cdots + \alpha_n x_n$. Define $m(x, B) = (\alpha_1 \cdots \alpha_n) \in D_{1 \times n}$. The map $h: V \to D_{1 \times n}$ given by $(x)h = m(x, B)$ is an isomorphism.

Proof

The proof is left as an exercise for the reader.

Proposition 5

Let ${}_D V$, ${}_D W$ be vector spaces, $B = \{x_1, \cdots, x_n\}$ a basis for V, $C = \{y_1, \cdots, y_m\}$ a basis for W, and $f: V \to W$ a homomorphism. Then

$$m((x)f, C) = m(x, B)m(f, B, C).$$

Proof

Let $m(x, B) = (\beta_1 \cdots \beta_n)$ and $m(f, B) = (\alpha_{ij})$. Then $x = \beta_1 x_1 + \cdots + \beta_n x_n$ and $(x_i)f = (\alpha_{i1} y_1 + \cdots + \alpha_{im} y_m)$, so

$$\begin{aligned}(x)f &= \left(\sum_{i=1}^{n} \beta_i x_i\right) f = \sum_{i=1}^{n} \beta_i (x_i) f \\ &= \sum_{i=1}^{n} \beta_i \left(\sum_{j=1}^{m} \alpha_{ij} y_j\right) = \sum_{j=1}^{m} \left(\sum_{i=1}^{n} \beta_i \alpha_{ij}\right) y_j\end{aligned}$$

and

$$\begin{aligned}m((x)f, C) &= \left(\sum_{i=1}^{n} \beta_i \alpha_{i1} \cdots \sum_{i=1}^{n} \beta_i \alpha_{im}\right) \\ &= (\beta_1 \cdots \beta_m) \begin{pmatrix} \alpha_{11} & \alpha_{12} & \cdots & \alpha_{1m} \\ \alpha_{21} & \alpha_{22} & \cdots & \alpha_{2m} \\ \vdots & & & \vdots \\ \alpha_{n1} & \alpha_{n2} & \cdots & \alpha_{nm} \end{pmatrix} \\ &= m(x, B)m(f, B, C).\end{aligned}$$

Proposition 5 tells us that, relative to fixed bases, matrices not only represent homomorphisms but may be used to compute the effect of a homomorphism on a given vector.

EXERCISES

1. Perform the indicated additions and multiplications.

(a)
$$\begin{pmatrix} 10 & -6 & 3 \\ -1 & 2 & 0 \\ 0 & 4 & 1 \end{pmatrix} + \begin{pmatrix} 3 & 1 & 2 \\ -4 & -1 & 2 \\ 1 & 0 & 1 \end{pmatrix}.$$

(b)
$$\begin{pmatrix} 10 & -6 & 3 \\ -1 & 2 & 0 \\ 0 & 4 & 1 \end{pmatrix} \cdot \begin{pmatrix} 3 & 1 & 2 \\ -4 & -1 & 2 \\ 1 & 0 & 1 \end{pmatrix}.$$

(c)
$$\begin{pmatrix} 5 & 0 & 2 \\ -1 & 3 & 1 \end{pmatrix} \cdot \begin{pmatrix} 2 & -1 & 4 \\ -6 & 10 & 0 \\ 1 & 0 & 1 \end{pmatrix}.$$

2. Prove Proposition 4.

3. Compute the following:

(a)
$$\begin{pmatrix} \alpha_{11} & 0 & 0 \\ 0 & \alpha_{22} & \alpha_{23} \\ 0 & \alpha_{32} & \alpha_{33} \end{pmatrix} \cdot \begin{pmatrix} \beta_{11} & 0 & 0 \\ 0 & \beta_{22} & \beta_{23} \\ 0 & \beta_{32} & \beta_{33} \end{pmatrix}.$$

(b)
$$\begin{pmatrix} \alpha_{11} & \alpha_{12} & 0 \\ \alpha_{21} & \alpha_{22} & 0 \\ \alpha_{31} & \alpha_{32} & 1 \end{pmatrix} \cdot \begin{pmatrix} \beta_{11} & \beta_{12} & 0 \\ \beta_{21} & \beta_{22} & 0 \\ \beta_{31} & \beta_{32} & 1 \end{pmatrix}.$$

*4. Let $A = (\alpha_{ij})$, $B = (\beta_{ij})$, $C = (\gamma_{ij}) \in D_n$ and suppose $A \cdot B = C$.

(a) Let

$$A' = \begin{pmatrix} \alpha_{22} & \cdots & \alpha_{2n} \\ \vdots & & \vdots \\ \alpha_{n2} & \cdots & \alpha_{nn} \end{pmatrix}, \qquad B' = \begin{pmatrix} \beta_{22} & \cdots & \beta_{2n} \\ \vdots & & \vdots \\ \beta_{n2} & \cdots & \beta_{nn} \end{pmatrix},$$

$$C' = \begin{pmatrix} \gamma_{22} & \cdots & \gamma_{2n} \\ \vdots & & \vdots \\ \gamma_{n2} & \cdots & \gamma_{nn} \end{pmatrix},$$

$\alpha_{i1} = \beta_{i1} = 0$ for $i \geq 2$ and $\alpha_{1j} = \beta_{1j} = 0$ for $j \geq 2$. Prove that $\gamma_{i1} = 0$ for $i \geq 2$, $\gamma_{1j} = 0$ for $j \geq 2$, $\gamma_{11} = \alpha_{11}\beta_{11}$, and $A'B' = C'$. [1.9]

(b) Let

$$A'' = \begin{pmatrix} \alpha_{11} & \cdots & \alpha_{1\,n-1} \\ \vdots & & \vdots \\ \alpha_{n-1\,1} & \cdots & \alpha_{n-1\,n-1} \end{pmatrix}, \quad B'' = \begin{pmatrix} \beta_{11} & \cdots & \beta_{1\,n-1} \\ \vdots & & \vdots \\ \beta_{n-1\,1} & \cdots & \beta_{n-1\,n-1} \end{pmatrix}$$

$$C'' = \begin{pmatrix} \gamma_{11} & \cdots & \gamma_{1\,n-1} \\ \vdots & & \vdots \\ \gamma_{n-1\,1} & \cdots & \gamma_{n-1\,n-1} \end{pmatrix},$$

$\alpha_{in} = \beta_{in} = 0$ for $i \leq n - 1$, $\alpha_{nn} = \beta_{nn} = 1$. Show that $A''B'' = C''$, $\gamma_{in} = 0$ for $i \leq n - 1$ and $\gamma_{nn} = 1$. [2.7]

5. Let $V = \mathbb{R}^3$, $B = \{e_1, e_2, e_3\}$ be the standard basis of $\mathbb{R}^3$, that is, $e_1 = (1, 0, 0)$, $e_2 = (0, 1, 0)$, $e_3 = (0, 0, 1)$, and $f:\mathbb{R}^3 \to \mathbb{R}^3$ the homomorphism such that

$$m(f, B) = \begin{pmatrix} 1 & 0 & 1 \\ 0 & 1 & 0 \\ 1 & 1 & 0 \end{pmatrix}.$$

Show that f is an isomorphism.

6. Let ${}_DV$ be a finite dimensional vector space and $f:V \to V$ a homomorphism. If for any bases B, B' of V, $m(f, B) = m(f, B')$, show that there exists $\lambda \in D$ such that $(x)f = \lambda x$ for all $x \in V$.

7. In $\mathbb{R}_4$ consider the elements

$$i = \begin{pmatrix} 0 & 1 & 0 & 0 \\ -1 & 0 & 0 & 0 \\ 0 & 0 & 0 & -1 \\ 0 & 0 & 1 & 0 \end{pmatrix}, \quad j = \begin{pmatrix} 0 & 0 & 1 & 0 \\ 0 & 0 & 0 & 1 \\ -1 & 0 & 0 & 0 \\ 0 & -1 & 0 & 0 \end{pmatrix},$$

$$k = \begin{pmatrix} 0 & 0 & 0 & 1 \\ 0 & 0 & -1 & 0 \\ 0 & 1 & 0 & 0 \\ -1 & 0 & 0 & 0 \end{pmatrix}.$$

(a) Show that $i^2 = j^2 = k^2 = -I$, and $ij = k$, $jk = i$, $ki = j$.
(b) For $\alpha \in \mathbb{R}$, and $(\alpha_{ij}) \in \mathbb{R}_4$, we denote by $\alpha \cdot (\alpha_{ij})$ the matrix whose ij entry is $\alpha\alpha_{ij}$. Show that the set $\mathcal{Q} = \{\alpha I + \beta i + \gamma j + \delta k : \alpha, \beta, \gamma, \delta \in \mathbb{R}\}$ is a division ring (but not a field) under matrix addition and multiplication. $\mathcal{Q}$ is called the *real quarternions*. [*Hint*: Show that the

inverse of $\alpha I + \beta i + \gamma j + \delta k$ is $\mu^{-1}(\alpha I - \beta i - \gamma j - \delta k)$, where $\mu = \alpha^2 + \beta^2 + \gamma^2 + \delta^2$. Notice the resemblance to the complex numbers.]

*8. Given ${}_{\mathbb{R}}V$ with basis $\{x_1, \cdots, x_n\} = B$, we define a new vector space $\hat{V}$ over $\mathbb{C}$ by

$$\hat{V} = \{x + yi : x, y \in V\}.$$

We define equality on the symbols $x + yi$ by

$$x + yi = x' + y'i \text{ if } x = x' \text{ and } y = y',$$

addition by

$$(x + yi) + (x' + y'i) = (x + x') + (y + y')i,$$

and scalar multiplication by

$$(\alpha + \beta i)(x + yi) = (\alpha x - \beta y) + (\alpha y + \beta x)i.$$

(a) Show that ${}_{\mathbb{C}}\hat{V}$ is a vector space. $\hat{V}$ is called the complexification of ${}_{\mathbb{R}}V$.
(b) Show that $\hat{B} = \{x_1 + 0i, \cdots, x_n + 0i\}$ is a basis for ${}_{\mathbb{C}}\hat{V}$.
(c) Let $f \in \text{Hom}_{\mathbb{R}}(V, V)$. Define $\hat{f}: \hat{V} \to \hat{V}$ by

$$(x + yi)\hat{f} = (x)f + (y)fi.$$

Show that $\hat{f} \in \text{Hom}_{\mathbb{C}}(\hat{V}, \hat{V})$.
(d) Letting $\hat{f}$ and $\hat{B}$ be as above, show that $m(\hat{f}, \hat{B}) = m(f, B)$. [3.5]

SECTION 1.9

ELEMENTARY MATRICES

In this section our attention will be restricted to square matrices, that is, elements of D_n. The portion of the development that can be extended to $D_{n \times m}$ will be discussed in the Exercises. Let I (or I_n) denote the matrix corresponding to the identity homomorphism.

DEFINITION 1 A matrix $(\alpha_{ij}) \in D_n$ is called *nonsingular* if there exists $(\beta_{ij}) \in D_n$ such that $(\alpha_{ij})(\beta_{ij}) = I$. (β_{ij}) is called the inverse of (α_{ij}) and is often written $(\alpha_{ij})^{-1}$.

Proposition 1

$(\alpha_{ij}) \in D_n$ is nonsingular $\Leftrightarrow$ for any vector space ${}_DV$ of dimension n and any basis B of V, the unique $f: V \to V$ such that $m(f, B) = (\alpha_{ij})$ is an isomorphism.

Proof

$\boxed{\Rightarrow}$: Suppose $(\alpha_{ij})(\beta_{ij}) = I$ and let $g: V \to V$ be such that $m(g, B) = (\beta_{ij})$. Then

$$I = m(f, B)m(g, B) = m(fg, B) \Rightarrow (x_i)fg = x_i \text{ for all } x_i \in B \Rightarrow g = f^{-1}$$

$\boxed{\Leftarrow}$: If $f: V \to V$ is an isomorphism, $f^{-1}: V \to V$ exists and so $I = m(ff^{-1}, B) = m(f, B)m(f^{-1}, B)$.

Corollary 1

If (α_{ij}), (β_{ij}) are nonsingular, then $(\alpha_{ij})(\beta_{ij})$ is nonsingular.

Properly speaking, the element $(\alpha_{ij})^{-1}$ as we have defined it is only a right inverse. Corollary 2 justifies our omission of the qualifying adjective. See Exercise 7 of Section 1.7 for a situation where a right inverse is not a left inverse.

Corollary 2

If $(\alpha_{ij})(\beta_{ij}) = I$, then $(\beta_{ij})(\alpha_{ij}) = I$.

The determination of the inverse of a matrix is often of importance, for example, in the simultaneous solution of equations. In solving n equations in n unknowns we have $\{\alpha_{ij} \in D: 1 \le i \le n, 1 \le j \le n\}$, $\lambda_1, \cdots, \lambda_n \in D$, and we want $\xi_1, \cdots, \xi_n \in D$ such that

$$\begin{array}{c} \alpha_{11}\xi_1 + \alpha_{12}\xi_2 + \cdots + \alpha_{1n}\xi_n = \lambda_1 \\ \alpha_{21}\xi_1 + \alpha_{22}\xi_2 + \cdots + \alpha_{2n}\xi_n = \lambda_2 \\ \vdots \qquad\qquad\qquad\qquad \vdots \\ \alpha_{n1}\xi_1 + \alpha_{n2}\xi_2 + \cdots + \alpha_{nn}\xi_n = \lambda_n . \end{array}$$

This may be rewritten as the matrix equation

$$(\alpha_{ij})\begin{pmatrix} \xi_1 \\ \vdots \\ \xi_n \end{pmatrix} = \begin{pmatrix} \lambda_1 \\ \vdots \\ \lambda_n \end{pmatrix}.$$

If we can find $(\alpha_{ij})^{-1}$, then we have

$$\begin{pmatrix} \xi_1 \\ \vdots \\ \xi_n \end{pmatrix} = (\alpha_{ij})^{-1}\begin{pmatrix} \lambda_1 \\ \vdots \\ \lambda_n \end{pmatrix}$$

and the equations are solved.

There is no really easy way to calculate the inverse of a nonsingular matrix. The method of elementary matrices developed in this section is about as efficient as any method and has other useful applications as well.

In solving systems of linear equations in high-school algebra, you learned that you could multiply through an equation by a constant, add a constant multiple of one equation to another, and change the order in which the equations were written, all without changing the simultaneous solution. Moreover, you learned that you could find the simultaneous solution by a systematic application of these techniques. Elementary matrices are the matrix analogue of these operations on systems of equations.

DEFINITION 2 The matrix $(\alpha_{ij}) \in D_n$ such that $\alpha_{kl} = 1$ and $\alpha_{ij} = 0$ for $i \neq k$ or $j \neq l$ will be denoted e_{kl}. If $(\alpha_{ij}) \in D_n$ and $\lambda \in D$, we shall write $\lambda(\alpha_{ij}) = (\lambda\alpha_{ij})$.

Proposition 2

$$e_{ij}e_{kl} = \begin{Bmatrix} (0) & \text{if} & j \neq k \\ e_{il} & \text{if} & j = k \end{Bmatrix}.$$

Proof

The proof is left to the reader.

DEFINITION 3 An *elementary matrix* (in D_n) is one of the following three types:

(a) $E(\lambda; i) = \lambda e_{ii} + \sum_{j=1, j \neq i}^{n} e_{jj}$, where $\lambda \neq 0$;
(b) $E(\lambda; i, j) = I + \lambda e_{ij}$, where $i \neq j$;
(c) $E(i, j) = I - e_{ii} - e_{jj} + e_{ij} + e_{ji}$.

In Exercise 2 you will verify that multiplying $(\alpha_{ij}) \in D_n$ on the left by $E(\lambda; i)$ leaves (α_{ij}) fixed except that the ith row is multiplied by λ, and that multiplying (α_{ij}) on the right by $E(\lambda; i)$ leaves (α_{ij}) fixed except that the ith column is multiplied by λ. Multiplying (α_{ij}) on the left by $E(\lambda; i, j)$ leaves all rows of (α_{ij}) fixed except the ith, which has λ times the jth row added to it. Multiplying (α_{ij}) on the right by $E(\lambda; i, j)$ adds λ times the ith column to the jth column. Multiplying on the left by $E(i, j)$ interchanges the ith and jth rows, and multiplying by $E(i, j)$ on the right interchanges the ith and jth columns.

As an example, let

$$A = \begin{pmatrix} 1 & 0 & 1 \\ 2 & 2 & 0 \\ -1 & 1 & 1 \end{pmatrix} \in \mathbb{R}_3.$$

Then

$$E(5; 2) \cdot A = \begin{pmatrix} 1 & 0 & 0 \\ 0 & 5 & 0 \\ 0 & 0 & 1 \end{pmatrix} \begin{pmatrix} 1 & 0 & 1 \\ 2 & 2 & 0 \\ -1 & 1 & 1 \end{pmatrix} = \begin{pmatrix} 1 & 0 & 1 \\ 10 & 10 & 0 \\ -1 & 1 & 1 \end{pmatrix}$$

and

$$A \cdot E(5; 2) = \begin{pmatrix} 1 & 0 & 1 \\ 2 & 2 & 0 \\ -1 & 1 & 1 \end{pmatrix} \begin{pmatrix} 1 & 0 & 0 \\ 0 & 5 & 0 \\ 0 & 0 & 1 \end{pmatrix} = \begin{pmatrix} 1 & 0 & 1 \\ 2 & 10 & 0 \\ -1 & 5 & 1 \end{pmatrix};$$

$$E(5; 1, 3) \cdot A = \begin{pmatrix} 1 & 0 & 5 \\ 0 & 1 & 0 \\ 0 & 0 & 1 \end{pmatrix} \begin{pmatrix} 1 & 0 & 1 \\ 2 & 2 & 0 \\ -1 & 1 & 1 \end{pmatrix} = \begin{pmatrix} -4 & 5 & 6 \\ 2 & 2 & 0 \\ -1 & 1 & 1 \end{pmatrix}$$

and

$$A \cdot E(5; 1, 3) = \begin{pmatrix} 1 & 0 & 1 \\ 2 & 2 & 0 \\ -1 & 1 & 1 \end{pmatrix} \begin{pmatrix} 1 & 0 & 5 \\ 0 & 1 & 0 \\ 0 & 0 & 1 \end{pmatrix} = \begin{pmatrix} 1 & 0 & 6 \\ 2 & 2 & 10 \\ -1 & 1 & -4 \end{pmatrix};$$

$$E(1, 2)A = \begin{pmatrix} 0 & 1 & 0 \\ 1 & 0 & 0 \\ 0 & 0 & 1 \end{pmatrix} \begin{pmatrix} 1 & 0 & 1 \\ 2 & 2 & 0 \\ -1 & 1 & 1 \end{pmatrix} = \begin{pmatrix} 2 & 2 & 0 \\ 1 & 0 & 1 \\ -1 & 1 & 1 \end{pmatrix}$$

and

$$A \cdot E(1, 2) = \begin{pmatrix} 1 & 0 & 1 \\ 2 & 2 & 0 \\ -1 & 1 & 1 \end{pmatrix} \begin{pmatrix} 0 & 1 & 0 \\ 1 & 0 & 0 \\ 0 & 0 & 1 \end{pmatrix} = \begin{pmatrix} 0 & 1 & 1 \\ 2 & 2 & 0 \\ 1 & -1 & 1 \end{pmatrix}.$$

Because of the way elementary matrices are defined it is easy to see not only that they have inverses, but what the inverses are.

Proposition 3

The elementary matrices are nonsingular.

Proof

$E(\lambda; i)E(\lambda^{-1}; i) = I$ (note that $\lambda \neq 0$ by definition), $E(\lambda; i, j)E(-\lambda; i, j) = I$, and $E(i, j)E(i, j) = I$.

Hence we know that a product of elementary matrices is nonsingular. It also happens that every nonsingular matrix is a product of elementary matrices. Ulterior motives that will shortly become apparent lead us to prove a somewhat more general result.

Theorem 1

Let $A \in D_n$, $A \neq (0)$. Denote by $J_k \in D_n$ the matrix $J_k = \sum_{i=1}^{k} e_{ii}$, $1 \leq k \leq n$. There exist elementary matrices $E_1, \cdots, E_r, F_1, \cdots, F_s \in D_n$ and k, $1 \leq k \leq n$ such that $A = F_1 \cdot F_2 \cdots F_s \cdot J_k \cdot E_r \cdots E_1$.

Proof

The proof is by induction on n. For $n = 1$ the result is obvious, for $A \in D_1 \Rightarrow A = E(\alpha; 1) = E(\alpha; 1) \cdot I = E(\alpha; 1)J_1 = E(\alpha; 1)J_1 E(1; 1)$.

Suppose the theorem is true for D_{n-1} and let $A \in D_n$. Then

$$(0) \neq A = \begin{pmatrix} \alpha_{11} & \alpha_{12} & \cdots & \alpha_{1n} \\ \alpha_{21} & \alpha_{22} & \cdots & \alpha_{2n} \\ \vdots & \vdots & & \vdots \\ \alpha_{n1} & \alpha_{n2} & \cdots & \alpha_{nn} \end{pmatrix} \Rightarrow \alpha_{ij} \neq 0 \quad \text{for some } i \text{ and } j.$$

The idea is to use type (c) elementary matrices to put a nonzero element in the (1, 1) position, then a type (a) elementary matrix to put a 1 in the (1, 1) position, and then type (b) elementary matrices to add suitable multiples of the first row to the other rows and of the first column to the other columns to get 0's everywhere in the first row and column except at the (1, 1) position. Writing this out explicitly, we have that there exist $\beta_{ij} \in D$, $2 \leq i \leq n$, $2 \leq j \leq n$ such that

$$\begin{pmatrix} 1 & 0 & \cdots & 0 \\ 0 & \beta_{22} & \cdots & \beta_{2n} \\ \vdots & \vdots & & \vdots \\ 0 & \beta_{n2} & \cdots & \beta_{nn} \end{pmatrix} = F'_{n+1} \cdot F'_n \cdot \cdots \cdot F'_1 \cdot A \cdot E'_1 \cdot E'_2 \cdot \cdots \cdot E'_n,$$

where $F'_1 = E(1, i)$, $E'_1 = E(1, j)$, i and j such that $\alpha_{ij} \neq 0$, $F'_2 = E(\alpha_{ij}{}^{-1}, 1)$. Also for $3 \leq k \leq n + 1$,

$$F'_k = E(-\alpha_{k-1,j}\,;\, k-1, 1),$$

except

$$F'_{i+1} = E(-\alpha_{1j}\,;\, i, 1),$$

since the 1st and ith rows have been interchanged. Similarly, for $2 \leq k \leq n$,

$$E'_k = E(-\alpha_{ij}{}^{-1}\alpha_{ik}\,;\, 1, k),$$

except

$$E'_j = E(-\alpha_{ij}{}^{-1}\alpha_{i1};\, 1, j),$$

since the 1st and jth columns have been interchanged.

Now by induction, if we put

$$B = \begin{pmatrix} \beta_{22} & \cdots & \beta_{2n} \\ \vdots & & \vdots \\ \beta_{n2} & \cdots & \beta_{nn} \end{pmatrix},$$

then there exist elementary matrices $F'_{n+2}, \cdots, F'_s, E'_{n+1}, \cdots, E'_r$, and some $J'_{k-1} \in D_{n-1}$ such that

$$B = F'_{n+2} \cdots F'_s J'_{k-1} E'_r \cdots E'_{n+1}.$$

Let

$$E_i = \begin{pmatrix} 1 & 0 & \cdots & 0 \\ 0 & & & \\ \vdots & & E'_i & \\ 0 & & & \end{pmatrix}, \quad J_k = \begin{pmatrix} 1 & 0 & \cdots & 0 \\ 0 & & & \\ \vdots & & J'_k & \\ 0 & & & \end{pmatrix}, \quad F_i = \begin{pmatrix} 1 & 0 & \cdots & 0 \\ 0 & & & \\ \vdots & & F'_i & \\ 0 & & & \end{pmatrix}.$$

Using Exercise 4, Section 1.8, we have

$$F'_{n-1} \cdots F'_1 \cdot A \cdot E'_1 \cdots E'_n = F_{n+2} \cdots F_s \cdot J_k \cdot E_r \cdots E_{n+1}.$$

Letting $F_i = (F'_i)^{-1}$, $E_j = (E'_j)^{-1}$, $1 \le i \le n+1$, $1 \le j \le n$, we have

$$A = F_1 \cdots F_{n+1} \cdot F_{n+2} \cdots F_s \cdot J_k \cdot E_r \cdots E_{n+1} \cdot E_n \cdots E_1.$$

Q.E.D.

Corollary

If $A \in D_n$ is nonsingular, then there exist elementary matrices $F_1, \cdots, F_p, E_1, \cdots, E_q$ such that

$$F_1 \cdots F_p A E_q \cdots E_1 = I \text{ and } A^{-1} = E_q \cdots E_1 \cdot F_1 \cdots F_p.$$

EXAMPLE Find the inverse of $\begin{pmatrix} 1 & 1 \\ 1 & -1 \end{pmatrix}$.

$$\begin{pmatrix} 1 & 0 \\ -1 & 1 \end{pmatrix}\begin{pmatrix} 1 & 1 \\ 1 & -1 \end{pmatrix}\begin{pmatrix} 1 & -1 \\ 0 & 1 \end{pmatrix} = \begin{pmatrix} 1 & 1 \\ 0 & -2 \end{pmatrix}\begin{pmatrix} 1 & -1 \\ 0 & 1 \end{pmatrix} = \begin{pmatrix} 1 & 0 \\ 0 & -2 \end{pmatrix},$$

$$\begin{pmatrix} 1 & 0 \\ 0 & -\frac{1}{2} \end{pmatrix}\begin{pmatrix} 1 & 0 \\ -1 & 1 \end{pmatrix}\begin{pmatrix} 1 & 1 \\ 1 & -1 \end{pmatrix}\begin{pmatrix} 1 & -1 \\ 0 & 1 \end{pmatrix} = \begin{pmatrix} 1 & 0 \\ 0 & 1 \end{pmatrix};$$

so

$$\begin{pmatrix} 1 & 1 \\ 1 & -1 \end{pmatrix}^{-1} = \begin{pmatrix} 1 & -1 \\ 0 & 1 \end{pmatrix}\begin{pmatrix} 1 & 0 \\ 0 & -\frac{1}{2} \end{pmatrix}\begin{pmatrix} 1 & 0 \\ -1 & 1 \end{pmatrix}$$

$$= \begin{pmatrix} 1 & \frac{1}{2} \\ 0 & -\frac{1}{2} \end{pmatrix}\begin{pmatrix} 1 & 0 \\ -1 & 1 \end{pmatrix}$$

$$= \begin{pmatrix} \frac{1}{2} & \frac{1}{2} \\ \frac{1}{2} & -\frac{1}{2} \end{pmatrix}.$$

This method, however, is still quite cumbersome if n is large. But we note that

$$E_q \cdots \cdot E_1 \cdot F_1 \cdot \cdots \cdot F_p A = I.$$

If we interpret the multiplication on the left by an elementary matrix as an operation on rows we have the following algorithm for finding inverses.

ALGORITHM Let A be a nonsingular matrix. Operate on its rows via interchange, multiplication by a constant, and adding a multiple of one row to another so as to get the identity matrix. If the operations applied to A to get I are applied to I they will yield A^{-1}.

EXAMPLE 1

$$A = \begin{pmatrix} 1 & 1 \\ 1 & -1 \end{pmatrix} \qquad I = \begin{pmatrix} 1 & 0 \\ 0 & 1 \end{pmatrix}.$$

Subtract 1st row from 2d:

$$\begin{pmatrix} 1 & 1 \\ 0 & -2 \end{pmatrix} \qquad \begin{pmatrix} 1 & 0 \\ -1 & 1 \end{pmatrix}.$$

Multiply 2d row by $-\frac{1}{2}$:

$$\begin{pmatrix} 1 & 1 \\ 0 & 1 \end{pmatrix} \qquad \begin{pmatrix} 1 & 0 \\ +\frac{1}{2} & -\frac{1}{2} \end{pmatrix}.$$

Subtract 2d row from 1st:

$$\begin{pmatrix} 1 & 0 \\ 0 & 1 \end{pmatrix} \qquad \begin{pmatrix} \frac{1}{2} & \frac{1}{2} \\ \frac{1}{2} & -\frac{1}{2} \end{pmatrix}.$$

EXAMPLE 2

(a)

$$\begin{pmatrix} 0 & 1 & 3 \\ 1 & 2 & 1 \\ 2 & 0 & 1 \end{pmatrix} \qquad \begin{pmatrix} 1 & 0 & 0 \\ 0 & 1 & 0 \\ 0 & 0 & 1 \end{pmatrix}$$

(b)

$$\begin{pmatrix} 1 & 2 & 1 \\ 0 & 1 & 3 \\ 2 & 0 & 1 \end{pmatrix} \qquad \begin{pmatrix} 0 & 1 & 0 \\ 1 & 0 & 0 \\ 0 & 0 & 1 \end{pmatrix}$$

(c)
$$\begin{pmatrix} 1 & 2 & 1 \\ 0 & 1 & 3 \\ 0 & -4 & -1 \end{pmatrix} \quad \begin{pmatrix} 0 & 1 & 0 \\ 1 & 0 & 0 \\ 0 & -2 & 1 \end{pmatrix}$$

(d)
$$\begin{pmatrix} 1 & 0 & -5 \\ 0 & 1 & 3 \\ 0 & -4 & -1 \end{pmatrix} \quad \begin{pmatrix} -2 & 1 & 0 \\ 1 & 0 & 0 \\ 0 & -2 & 1 \end{pmatrix}$$

(e)
$$\begin{pmatrix} 1 & 0 & -5 \\ 0 & 1 & 3 \\ 0 & 0 & 11 \end{pmatrix} \quad \begin{pmatrix} -2 & 1 & 0 \\ 1 & 0 & 0 \\ 4 & -2 & 1 \end{pmatrix}$$

(f)
$$\begin{pmatrix} 1 & 0 & -5 \\ 0 & 1 & 3 \\ 0 & 0 & 1 \end{pmatrix} \quad \begin{pmatrix} -2 & 1 & 0 \\ 1 & 0 & 0 \\ \frac{4}{11} & -\frac{2}{11} & \frac{1}{11} \end{pmatrix}$$

(g)
$$\begin{pmatrix} 1 & 0 & 0 \\ 0 & 1 & 3 \\ 0 & 0 & 1 \end{pmatrix} \quad \begin{pmatrix} -\frac{2}{11} & \frac{1}{11} & \frac{5}{11} \\ 1 & 0 & 0 \\ \frac{4}{11} & -\frac{2}{11} & \frac{1}{11} \end{pmatrix}$$

(h)
$$\begin{pmatrix} 1 & 0 & 0 \\ 0 & 1 & 0 \\ 0 & 0 & 1 \end{pmatrix} \quad \begin{pmatrix} -\frac{2}{11} & \frac{1}{11} & \frac{5}{11} \\ -\frac{1}{11} & \frac{6}{11} & -\frac{3}{11} \\ \frac{4}{11} & -\frac{2}{11} & \frac{1}{11} \end{pmatrix}.$$

As a check we see that

$$\begin{pmatrix} 0 & 1 & 3 \\ 1 & 2 & 1 \\ 2 & 0 & 1 \end{pmatrix} \begin{pmatrix} -\frac{2}{11} & \frac{1}{11} & \frac{5}{11} \\ -\frac{1}{11} & \frac{6}{11} & -\frac{3}{11} \\ \frac{4}{11} & -\frac{2}{11} & \frac{1}{11} \end{pmatrix} = \begin{pmatrix} 1 & 0 & 0 \\ 0 & 1 & 0 \\ 0 & 0 & 1 \end{pmatrix}.$$

Clearly it is possible to add to the efficiency of this process. For example, steps (g) and (h) can be carried out together as can steps (c) and (d). Indeed, we only wrote them out individually to aid you in following the computation.

DEFINITION 4 Let $A = (\alpha_{ij}) \in D_{n \times m}$.

(a) $A^T = (\beta_{ij}) \in D_{m \times n}$ where $\beta_{ij} = \alpha_{ji}$. A^T is called the *transpose* of A.

(b) The *row rank of A* is the maximal number of linearly independent rows of A where the rows of A are identified with elements of D^m.

(c) The *column rank* of A is the maximal number of linearly independent columns of A where the columns of A are identified with elements of D^n.

Proposition 4

Let $A \in D_{n \times m}$, $B \in D_{m \times r}$ and $E \in D_n$. Then:

(**a**) if E is an elementary matrix then E^T is also;
(**b**) $(A^T)^T = A$;
(**c**) column rank of A = row rank of A^T;
(**d**) moreover, if D is a field then $(AB)^T = B^T A^T$.

Proof

The proof is left as an exercise for the reader.

Proposition 5

Let $A, A' \in D_n$, A nonsingular. Then row rank (A') = row rank (AA') = row rank $(A'A)$.

Proof

Let V be a vector space of dimension n, $B = \{x_1, \cdots, x_n\}$ a basis of V, $f, g: V \to V$ homomorphisms such that $m(f, B) = A$, $m(g, B) = A'$. Observe that row rank (A') = dim (Im (g)), since any set of linearly independent rows of A' corresponds to a linearly independent set of elements in the image of g, and if $A' = (\alpha'_{ij})$,

$$\text{Im}\,(g) = \langle(\alpha'_{11}x_1 + \cdots + \alpha'_{1n}x_n), \cdots, (\alpha'_{n1}x_1 + \cdots + \alpha'_{nn}x_n)\rangle.$$

A nonsingular $\Rightarrow f$ is an isomorphism $\Rightarrow$

$$\dim\,(\text{Im}\,(g)) = \dim\,(\text{Im}\,(fg)) = \dim\,(\text{Im}\,(gf)),$$

and the result follows.

Proposition 6

If $A \in D_n$, D a field, then row rank (A) = column rank (A).

Proof

We know $A = F_p \cdots F_1 \cdot J_k \cdot E_1 \cdot \cdots \cdot E_s$, where F_i, E_i elementary and J_k is as previously defined. Then row rank (A) = row rank $(J_k) = k$. Hence $A^T = E_s^T \cdots E_1^T J_k^T F_1^T \cdots F_p^T$, $J_k^T = J_k$ and column rank (A) = row rank $(A^T) = k$ = row rank (A).

EXERCISES

1. Find the multiplicative inverses of

$$\begin{pmatrix} 3 & 1 & 1 \\ 1 & 2 & 1 \\ -3 & 0 & 3 \end{pmatrix} \quad \text{and} \quad \begin{pmatrix} 1 & 0 & 1 & 0 \\ 0 & 1 & 0 & 1 \\ 1 & 1 & 1 & 0 \\ 0 & 0 & 1 & 1 \end{pmatrix}.$$

2. Find $\xi_1, \xi_2, \xi_3, \xi_4 \in \mathbb{R}$ such that $\xi_1 + \xi_3 = 5$; $\xi_2 + \xi_4 = -1$; $\xi_1 + \xi_2 + \xi_3 = 0$; $\xi_3 + \xi_4 = 7$.

3. The example following the corollary to Theorem 1 can be used to give a decomposition of $\begin{pmatrix} 1 & 1 \\ 1 & -1 \end{pmatrix}$ of the type guaranteed by the theorem. Find a second decomposition, thus showing that the decomposition of an $n \times n$ matrix into a product of elementary matrices and a matrix of the form J_k is not unique.

4. Prove an analogue to Theorem 1 for $D_{n \times m}$. Let $J_k = \sum_{i=1}^{k} e_{ii} \in D_{n \times m}$ and show that if $A \in D_{n \times m}$ there exist elementary matrices $F_1, \cdots, F_p \in D_n$ and elementary matrices $E_1, \cdots, E_s \in D_m$ such that $A = F_p \cdot \cdots \cdot F_1 J_k E_1 \cdot \cdots \cdot E_s$ for some k. [*Hint*: First prove that the result is true for $D_{1 \times m}$ for all $m \geq 1$, and then show that if it is true for $D_{n-1 \times m}$ it is true for $D_{n \times m}$.] As a corollary, deduce that the row rank equals the column rank for an element of $D_{n \times m}$, provided D is a field.

*5. An *upper* (*lower*) *triangular* matrix in D_n is a matrix (α_{ij}) such that $\alpha_{ij} = 0$ when $i > j$ (respectively $i < j$). Show that if (α_{ij}) is an upper triangular matrix, and $\alpha_{ii} = 1$, $i = 1, 2, \cdots, n$, then we can write

$$(\alpha_{ij}) = E(\alpha_{n-1,n}; n-1, n)E(\alpha_{n-2,n-1}; n-2, n-1) \cdots E(\alpha_{1n}; 1, n),$$

a product of elementary matrices of the form $E(\lambda; i, j)$. [3.7]

SECTION 1.10
DUAL SPACES

We have already shown (Section 1.8, Proposition 2) that if ${}_DV$, ${}_DW$ are vector spaces, $f, g: V \to W$ homomorphisms, then $f + g: V \to W$ defined by $(x)(f + g) = (x)f + (x)g$ is a homomorphism. We can extend this to the following:

Proposition 1

Let ${}_DV$, ${}_DW$ be vector spaces. Then under the binary operation "+," $\operatorname{Hom}_D(V, W)$ is an Abelian group.

Proof

Let $f, g, h \in \operatorname{Hom}_D(V, W)$. Then for $x \in V$,

$$\begin{aligned}(x)((f+g)+h) &= (x)(f+g) + (x)h\\ &= ((x)f + (x)g) + (x)h\\ &= (x)f + ((x)g + (x)h)\\ &= (x)f + (x)(g+h)\\ &= (x)(f + (g+h)),\end{aligned}$$

so $(f+g)+h = f+(g+h)$ and "+" is associative.

We leave it to the reader to check that the map $0: V \to W$ defined by $(x)0 = O$ for all $x \in V$, is in $\operatorname{Hom}_D(V, W)$ and for any $f \in \operatorname{Hom}_D(V, W)$, $f + 0 = f$. Also if $f \in \operatorname{Hom}_D(V, W)$ the map $-f: V \to W$ defined by $(x)(-f) = -(x)f$ is in $\operatorname{Hom}_D(V, W)$ and $f + (-f) = 0$. Moreover, $f + g = g + f$ for all $f, g \in \operatorname{Hom}_D(V, W)$.

$\operatorname{Hom}_D(V, W)$ cannot, in general, be made into a vector space, for suppose $f \in \operatorname{Hom}_D(V, W)$, $\alpha \in D$ and we wish to define the product αf. If $\operatorname{Hom}_D(V, W)$ is to be a vector space αf must be in $\operatorname{Hom}_D(V, W)$. If $x \in V$, $(x)(\alpha f) \in W$. A possible definition is $(x)(\alpha f) = \alpha(x)f$, but then $(\beta x)(\alpha f) = \alpha(\beta x)f = \alpha\beta(x)f$ and $(\beta x)(\alpha f) = \beta(x)(\alpha f) = \beta\alpha(x)f \Rightarrow \alpha\beta = \beta\alpha$ for all $\alpha, \beta \in D$, so D would be a field. There is, however, a useful special situation in which a vector space is obtained.

Proposition 2

Let D be a division ring, ${}_DV$ a vector space. ${}_DD(={}_DD^1)$ is a vector space, $\operatorname{Hom}_D(D, V)$ is a vector space, and $V \approx \operatorname{Hom}_D(D, V)$.

Proof

We know $\operatorname{Hom}_D(D, V)$ is an Abelian group under addition of homomorphisms. If $f \in \operatorname{Hom}_D(D, V)$, $\alpha \in D$, we define $\alpha f: D \to V$ by, for $\beta \in D$, $(\beta)(\alpha f) = (\beta\alpha)f$. αf is a homomorphism, for if $\beta_1, \beta_2 \in D$ then $(\beta_1 + \beta_2)(\alpha f) = (\beta_1\alpha + \beta_2\alpha)f = (\beta_1\alpha)f + (\beta_2\alpha)f = (\beta_1)(\alpha f) + (\beta_2)(\alpha f)$ and $(\beta_1\beta_2)\ (\alpha f) = (\beta_1\beta_2\alpha)f = \beta_1(\beta_2\alpha)f = \beta_1(\beta_2)(\alpha f)$. That this scalar multiplication is associative, distributive, and unitary is left to the reader.

If $x \in V$ define $f_x : D \to V$ by $(\alpha)f_x = \alpha x$. For $\alpha, \beta \in D$,

$$(\beta + \alpha)f_x = (\beta + \alpha)x = \beta x + \alpha x = (\beta)f_x + (\alpha)f_x \quad \text{and}$$
$$(\beta\alpha)f_x = (\beta\alpha)x = \beta(\alpha x) = \beta(\alpha)f_x,$$

so $f_x \in \operatorname{Hom}_D(D, V)$. We define $h : V \to \operatorname{Hom}_D(D, V)$ by $(x)h = f_x$ and claim that h is the required isomorphism. If $x, y \in V$, $\alpha \in D$ then

$$(\alpha)(f_x + f_y) = (\alpha)f_x + (\alpha)f_y = \alpha x + \alpha y = \alpha(x + y) = (\alpha)f_{x+y},$$

so $f_x + f_y = f_{x+y}$ and therefore $(x + y)h = (x)h + (y)h$. If $\alpha, \beta \in D$, $x \in V$, $(\alpha)f_{\beta x} = \alpha(\beta x) = (\alpha\beta)x = (\alpha)(\beta f_x) \Rightarrow f_{\beta x} = \beta f_x$ and $(\beta x)h = \beta(x)h$ for all $\beta \in D$, $x \in V$. Now that we know h is a homomorphism we look at the kernel. If $(x)h = 0$, that is, $f_x = 0$ then $(\alpha)f_x = \alpha x = O$ for all $\alpha \in D$; in particular $O = (1)f_x = x$ and so h is a monomorphism. If $f \in \operatorname{Hom}_D(D, V)$ let $x = (1)f$. Then $(\alpha)f = \alpha(1)f = \alpha x$ so $f = f_x = (x)h$ and h is an epimorphism.

Remark: In Section 1.8 we defined, for $x \in V$ and $B = \{x_1, \cdots, x_n\}$ a basis of V, the matrix $m(x, B)$. If we take $\{1\}$ as a basis for ${}_DD$ we see that $m(x, B) = m(f_x, \{1\}, B)$.

Proposition 3

Let ${}_DV$ be a vector space. For $f \in \operatorname{Hom}_D(V, D)$, $\alpha \in D$, define $f\alpha : V \to D$ by $(x)(f\alpha) = ((x)f)\alpha$ for all $x \in V$. Then:

(a) $f\alpha \in \operatorname{Hom}_D(V, D)$;
(b) if $\alpha, \beta \in D$, $(f\alpha)\beta = f(\alpha\beta)$;
(c) $f \cdot 1 = f$;
(d) if $\alpha, \beta \in D$, $f(\alpha + \beta) = f\alpha + f\beta$.

Proof

Note first that if $x \in V$ and $f \in \operatorname{Hom}_D(V, D)$, then $(x)f \in D$ so $(x)f \cdot \alpha$ is a product in D and the definition makes sense.

(a) Let $x, y \in V$. Then

$$\begin{aligned}(x + y)(f\alpha) &= ((x + y)f)\alpha = ((x)f + (y)f)\alpha \\ &= ((x)f)\alpha + ((y)f)\alpha \\ &= (x)(f\alpha) + (y)(f\alpha).\end{aligned}$$

If $\gamma \in D$, then

$$(\gamma x)(f\alpha) = ((\gamma x)f)\alpha = (\gamma(x)f)\alpha = \gamma(((x)f)\alpha) = \gamma(x)(f\alpha).$$

(b) For $x \in V$,

$$(x)((f\alpha)\beta) = ((x)(f\alpha))\beta = ((x)f \cdot \alpha)\beta = ((x)f)\alpha\beta = (x)(f(\alpha\beta)).$$

(c) $(x)(f \cdot 1) = (x)f \cdot 1 = (x)f.$

(d) $(x)(f(\alpha + \beta)) = (x)f(\alpha + \beta) = ((x)f)\alpha + ((x)f)\beta$
$$= (x)(f\alpha) + (x)(f\beta) = (x)(f\alpha + f\beta).$$

We observe that the facts that Hom $_D(V, D)$ is an Abelian group, D is a division ring, and Proposition 3 are very like an assertion that Hom $_D(V, D)$ is a vector space, the only difficulty being that scalar multiplication is defined on the right instead of the left. This leads to the following definition.

DEFINITION 1 A *right vector space* consists of

(a) an additively written Abelian group V,
(b) a division ring D,
(c) a map $V \times D \to V$ where the image of (x, α) is denoted $x\alpha$ satisfying
- **(i)** $x(\alpha\beta) = (x\alpha)\beta$,
- **(ii)** $x \cdot 1 = x$,
- **(iii)** $(x + y)\alpha = x\alpha + y\alpha$,
- **(iv)** $x(\alpha + \beta) = x\alpha + x\beta$.

We shall denote this by V_D.

What we have until now called vector spaces will be called *left vector spaces*. The unmodified phrase "vector space" will mean left vector space. Clearly there is a 1-1 correspondence between theorems about left vector spaces and theorems about right vector spaces. Homomorphisms between right vector spaces will be written on the left, that is, if $f: V_D \to W_D$ and $x \in V$ we will write $f(x)$. Note that if $g: W_D \to U_D$ the composite of f first and then g is gf and not fg.

DEFINITION 2 If $_DV$ is a left vector space, the right vector space Hom $_D(V, D)$ is denoted V^* and called *the dual space of* V.

We remark here that a right vector space W_D has a dual space $W^* =$ Hom $_D(W, D)$ that is a left vector space.

If $\dim(V) = n$, $B = \{x_1, \cdots, x_n\}$ a basis of V, $\{1\}$ a basis of D, $x \in V$, and $f \in V^*$, then $(x)f = m(x, B)m(f, B, \{1\})$ (we have identified $(x)f$ with $m((x)f, \{1\})$ in D). That is, if we identify V with $D_{1 \times n}$ we may identify V^* with $D_{n \times 1}$ and the operation of V^* on V with matrix multiplication. If you let $\{e_1, \cdots, e_n\} \in D_{1 \times n}$ be the basis such that e_i has a 1 in the ith position and 0's elsewhere and $e_1{}^T, \cdots, e_n{}^T \in D_{n \times 1}$ be a basis, you will note that

$$e_i e_j{}^T = \begin{Bmatrix} 0 & \text{if } i \neq j \\ 1 & \text{if } i = j \end{Bmatrix}.$$

More generally, we have the following proposition:

Proposition 4 (Dual Basis Lemma)

Let ${}_DV$ be a vector space and $B = \{x_1, \cdots, x_n\}$ a basis for V. Then there exists a basis $C = \{f_1, \cdots, f_n\}$ of V^* such that $(x_i)f_j = 0$ if $i \neq j$ and $(x_i)f_i = 1$, $1 \leq i \leq n$. In particular, $\dim(V) = \dim(V^*)$.

Proof

For $i = 1, \cdots, n$ let $f_i : V \to D$ be the unique homomorphism such that $(x_j)f_i = 0, j \neq i$, and $(x_i)f_i = 1$. If $f \in V^*$ let $(x_i)f = \alpha_i$. Then

$$(x_j)\left(\sum_{j=1}^{n} f_j \alpha_j\right) = \alpha_j = (x_j)f$$

for $j = 1, \cdots, n$ which implies $f = \sum_{j=1}^{n} f_j \alpha_j$, and hence $\langle f_1, \cdots, f_n \rangle = V^*$. If $\beta_1, \cdots, \beta_n \in D$ and $0 = f_1\beta_1 + \cdots + f_n\beta_n$, then

$$0 = (x_i)0 = (x_i)\left(\sum_{j=1}^{n} f_j \beta_j\right) = \beta_i,\ 1 \leq i \leq n,$$

so, $\{f_1, \cdots, f_n\}$ is linearly independent.

Note that since V is a left vector space and V^* is a right vector space, $\dim(V) = \dim(V^*)$ says nothing about an isomorphism between V and V^*. However, if we let $V^{**} = (V^*)^*$ then V^{**} is a left vector space and $\dim(V) = \dim(V^*) = \dim(V^{**})$ does imply an isomorphism from V to V^{**}. The following proof gives an alternative proof of this fact.

Proposition 5

Let ${}_DV$ be a finite dimensional vector space. If $x \in V$, $f \in V^*$, define $\hat{x}: V^* \to D$ by $\hat{x}(f) = (x)f$. Then:

(a) $\hat{x} \in V^{**}$;
(b) the map $\phi: V \to V^{**}$ defined by $(x)\phi = \hat{x}$ is an isomorphism.

Proof

(a) We must show that $\hat{x}$ is a homomorphism. If $f, g \in V^*$, then

$$\begin{aligned}\hat{x}(f+g) &= (x)(f+g)\\ &= (x)f + (x)g\\ &= \hat{x}(f) + \hat{x}(g).\end{aligned}$$

And if $\alpha \in D$, then $\hat{x}(f\alpha) = (x)(f\alpha) = ((x)f)\alpha = \hat{x}(f)\alpha$.

(b) To show $(x + y)\phi = (x)\phi + (y)\phi$ let $f \in V^*$. Then

$$\begin{aligned}(x + y)\phi(f) &= \widehat{x + y}(f)\\ &= (x + y)f\\ &= (x)f + (y)f\\ &= \hat{x}(f) + \hat{y}(f) = (x)\phi(f) + (y)\phi(f).\end{aligned}$$

If $\alpha \in D$ we have

$$\begin{aligned}(\alpha x)\phi(f) &= \widehat{\alpha x}(f) = (\alpha x)f\\ &= \alpha(x)f = \alpha\hat{x}(f)\\ &= \alpha(x)\phi(f) \text{ so } (\alpha x)\phi = \alpha(x)\phi.\end{aligned}$$

If $x \neq O$ there exist $x_2, \cdots, x_n \in V$ such that $\{x, x_2, \cdots, x_n\}$ is a basis and an $f \in V^*$ such that $(x)f = 1$, $(x_i)f = 0$, $2 \leq i \leq n$. Then $(x)\phi(f) = (x)f = 1 \Rightarrow (x)\phi \neq 0 \Rightarrow x \notin \operatorname{Ker}(\phi)$, so $O = \operatorname{Ker}(\phi)$. ϕ is an epimorphism since $\dim V = \dim(V^{**})$.

EXERCISES

1. Let $V = \mathbb{R}^4$, $H = \langle (1, 0, 0, 1), (1, 0, 1, 0) \rangle$. Find $\{f \in V^* : (x)f = 0 \text{ for all } x \in H\}$ and show that it is a subspace.

2. If V is a vector space and $H < V$, the *annihilator of H*, denoted Ann (H), is defined by

$$\operatorname{Ann}(H) = \{f \in V^* : (x)f = 0 \text{ for all } x \in H\}.$$

   (a) Prove that Ann $(H) < V^*$.
   (b) If $\dim(V) = n$ use the dual basis lemma to prove

$$\dim(\operatorname{Ann} H) = \dim(V) - \dim(H).$$

   (c) If $H < V$ and $K < V$ prove that

$$\operatorname{Ann}(H + K) = \operatorname{Ann}(H) \cap \operatorname{Ann}(K)$$

   and

$$\operatorname{Ann}(H \cap K) = \operatorname{Ann}(H) + \operatorname{Ann}(K).$$

3. (a) If $h: {}_D D \to {}_D D$ is a homomorphism show that there exists an $\alpha \in D$ such that $(\beta)h = \beta\alpha$ for all $\beta \in D$.
   (b) Let $f, g \in V^*$. Using part (a) and the fundamental theorem of homomorphisms show that $\operatorname{Ker}(f) = \operatorname{Ker}(g) \Leftrightarrow$ there exists $\alpha \in D$, $\alpha \neq 0$, such that $f = g\alpha$.

4. If D is a field show that $\operatorname{Hom}_D(V, W)$ is a vector space.

SECTION 1.11
ELEMENTARY PROPERTIES OF GROUPS

Throughout this chapter we have seen examples of groups occurring naturally in connection with vector spaces and homomorphisms. Indeed, vector spaces themselves are groups. This final section is intended to develop some rudiments of group theory that will be useful in the future. Where the present results are exact parallels of previous results, the proofs will be omitted. In case refinements must be made, we provide the necessary transitional remarks. The reader is urged to fill in the missing details, thereby reviewing for himself many of the techniques used in this chapter.

Recall that a *group* is a nonempty set G together with an associative binary operation such that there exists an element $e \in G$ called the *identity*, and, for each $g \in G$, there exists $g^{-1} \in G$ called the *inverse of* g satisfying $g \cdot e = e \cdot g = g$, and $g \cdot g^{-1} = g^{-1} \cdot g = e$.

Proposition 1

Let G be a group. Then:

(a) e is unique;
(b) g^{-1} is unique for each $g \in G$;
(c) $g^{-1} = h^{-1} \Leftrightarrow g = h$;
(d) $g \cdot h = g \cdot h' \Rightarrow h = h'$ for all $g, h, h' \in G$.

In practice it is convenient to use a variety of symbols for the binary operation on a group. In doing so, tradition demands additional adaptations. In particular, when we use "+" for the operation, we shall write $-g$ for the inverse of g and 0 for the identity.

DEFINITIONS *A* (*group*) *homomorphism* from a group G into a group G' is a function $f{:}G \to G'$ such that

$$(g_1 \cdot g_2)f = (g_1)f \cdot (g_2)f, \quad \forall g_1, g_2 \in G.$$

As usual, it is understood that the operation on the left-hand side of the equation means the binary operation on G, whereas the right-hand side of the equation refers to the binary operation on G'. We define and denote the *kernel of* f and the *image of* f respectively by $\text{Ker}(f) = \{g \in G{:}(g)f = e'$, the identity of $G'\}$; $\text{Im}(f) = \{g' \in G'{:}g' = (g)f$, for some $g \in G\}$. We say f is a *monomorphism* if $\text{Ker}(f) = e$. *Epimorphism* and *isomorphism* are defined as before. *Subgroups* have been defined in the exercises, and if H is a subgroup of G, we write $H < G$. Following Sections 1.2 and 1.5, we may assert:

Proposition 2

Let G be a group.

(a) Let $\varnothing \neq H \subseteq G$. Then $H < G \Leftrightarrow h_1 h_2^{-1} \in H, \forall h_1, h_2 \in H$.
(b) The intersection of a collection of subgroups of G is a subgroup of G.

Proposition 3

Let G and G' be groups, and $f:G \to G'$ a homomorphism.

(a) $(e)f = e'$, where e, e' are the respective identities of G, G'.
(b) $\text{Ker}(f) < G$.
(c) $\text{Im}(f) < G'$.
(d) f is a monomorphism $\Leftrightarrow f$ is 1-1.

Proposition 4

Let G, G', G'' be groups, and $f:G \to G', f':G' \to G''$ be homomorphisms.

(a) If f, f' are monomorphisms, then ff' is a monomorphism.
(b) If f, f' are epimorphisms, then ff' is an epimorphism.
(c) If f, f' are isomorphisms, then ff' is an isomorphism.

Continuing the pattern we define, for $H < G$, a *(left) coset of G relative to H* to be any subset of the form

$$g \cdot H = \{g \cdot h : h \in H\}.$$

Proposition 5

Let $H < G$. Then for $g_1, g_2 \in G$, the following are equivalent.

(a) $(g_1 \cdot H) \cap (g_2 \cdot H) \neq \varnothing$.
(b) $g_2^{-1} g_1 \in H$.
(c) $g_1 \cdot H = g_2 \cdot H$.

One would naturally expect now to define G/H, make it into a group, and prove the fundamental theorems of homomorphisms. At this point, however, the analogy between vector spaces and arbitrary groups becomes more delicate.

Let us put $G/H = \{g \cdot H : g \in G\}$. Suppose we try to define a binary operation on G/H by

$$(g_1 \cdot H) \cdot (g_2 \cdot H) = (g_1 \cdot g_2) \cdot H,$$

corresponding to the vector-space definition. As before, we must guarantee that this operation is meaningful, or *well-defined*. Specifically, since we have

$(g_1 \cdot h) \cdot H = g_1 \cdot H$, for any $h \in H$ (Proposition 5), it should follow that $((g_1 \cdot h) \cdot H) \cdot (g_2 \cdot H) = (g_1 \cdot h \cdot g_2) \cdot H$, that is, $(g_1 \cdot h \cdot g_2) \cdot H = (g_1 \cdot g_2) \cdot H$. The latter, however, implies that

$$\begin{aligned} & g_1 \cdot h \cdot g_2 = g_1 \cdot g_2 \cdot h', \text{ for some } h' \in H \\ \Leftrightarrow\ & h \cdot g_2 = g_2 \cdot h' \\ \Leftrightarrow\ & g_2^{-1} \cdot h \cdot g_2 = h', \text{ for some } h' \in H. \end{aligned}$$

In other words, if the "multiplication" is to be well defined in G/H, then the following condition must be satisfied:

$$\forall h \in H, \forall g \in G,\ g^{-1} \cdot h \cdot g \in H,$$

or briefly, $g^{-1} \cdot H \cdot g \subseteq H$. More importantly, this condition is *sufficient*, as the reader can verify by essentially reversing the above argument.

DEFINITION A subgroup H of a group G is called a *normal subgroup of* G if $g^{-1}Hg \subseteq H$, $\forall g \in G$. We write $H \lhd G$.

EXAMPLES

1. If G is a commutative group, then any subgroup is normal.
2. G and $\{e\}$ are normal subgroups of any group G.
3. For any vector space ${}_DV$, where D is a field, the automorphisms $\{f_\alpha : \alpha \in D\}$ form a normal subgroup of $GL(V)$, where we define $(x)f_\alpha = \alpha x$, $x \in V$.
4. More generally, if $H < G$, and $hg = gh$, $\forall g \in G$, and $\forall h \in H$ (as in Example 3), then $H \lhd G$.
5. The subgroup S of shear transformations of $\mathbb{R}^2$ (see Section 1.5, Exercise 6) is *not* a normal subgroup of $GL(\mathbb{R}^2)$, since if we define $g \in GL(\mathbb{R}^2)$ by $((1, 0))g = (0, 1)$ and $((0, 1))g = (1, 0)$, then $g^{-1} = g$, and $g^{-1}hg \notin S$ in general, when $h \in S$. (Check this computation.)

We now assert the fundamental theorems of group homomorphisms.

Proposition 6

Let $H \lhd G$. The collection G/H of left cosets of G relative to H is a group under the operation

$$(g_1 \cdot H) \cdot (g_2 \cdot H) = (g_1 g_2) \cdot H.$$

Furthermore, the function $\pi : G \to G/H$, given by $(g)\pi = g \cdot H$, is an epimorphism, and is called the *natural homomorphism* of G onto G/H.

Proposition 7

Let $f:G \to G'$ be an epimorphism with kernel K. Then $K \lhd G$, and there exists an isomorphism $\bar{f}:G/K \to G'$ such that $f = \pi\bar{f}$. Moreover, there is a 1-1 correspondence between the collection of subgroups of G that contain K and the collection of all subgroups of G'. This correspondence preserves inclusion and intersection, and it sends normal subgroups to normal subgroups.

As an application of the foregoing ideas, we outline the construction of a finite field. Let $\mathbb{Z}$ be the set of all integers (positive, negative, and zero) with the usual binary operations of addition and multiplication. Let n be a fixed positive integer, and put $H = \{m \in \mathbb{Z}:n \text{ divides } m\}$. Certainly H is a (normal) subgroup of the group $\mathbb{Z}$ under addition. Hence $\mathbb{Z}/H = \{x + H:x \in \mathbb{Z}\}$ becomes a group when we define $(x + H) + (y + H) = (x + y) + H$.

For each coset $x + H$ we assume that the representative x is the least nonnegative integer in $x + H$. By the "Division Algorithm," we see that there are precisely n distinct cosets, $0 + H, 1 + H, \cdots, (n - 1) + H$.

Our aim is to define a second binary operation, "multiplication," on $\mathbb{Z}/H$ so that $\mathbb{Z}/H$ satisfies the axioms of a field. In general this cannot be done; however, in case $n = p$, a *prime* integer, we can prove the following crucial result.

Lemma 1

Let p be a prime positive integer, and let $H = \{m \in \mathbb{Z}:p \text{ divides } m\}$. For any $x \in \mathbb{Z}$, $x > 0$, and x not divisible by p, there exists a $y \in \mathbb{Z}$ such that $yx + H = 1 + H$.

Proof

We assume the fact from elementary number theory that if p divides a product of two integers, then p must divide one of them. Now let x be given, not divisible by p, and consider the collection $1 \cdot x, 2 \cdot x, \cdots, (p - 1)x$ of integers. If $1 \leq a < b \leq p - 1$, then $0 < b - a < p - 1$, so that p does not divide $b - a$. Hence p cannot divide $bx - ax = (b - a)x$. In other words, using Proposition 5, the $p - 1$ cosets $1 \cdot x + H, 2 \cdot x + H, \cdots, (p - 1)x + H$ are distinct elements of $\mathbb{Z}/H$. Since none of these cosets is $0 + H$, one of them, say $y \cdot x + H$, must be $1 + H$. Q.E.D.

Using Lemma 1 (and its proof) to show the existence (and uniqueness) of multiplicative inverses, the reader should have no problem verifying the remaining axioms and showing multiplication is well defined to establish:

Proposition 8

Let p be a prime positive integer, and let $H = \{m \in \mathbb{Z} : p \text{ divides } m\}$. Then $\mathbb{Z}/H$ becomes a field when we define addition and multiplication by

$$(x + H) + (y + H) = (x + y) + H, \ (x + H)(y + H) = xy + H.$$

DEFINITION The finite field described in Proposition 8 is called the field of *integers modulo p*, and is denoted $\mathbb{Z}_p$.

We conclude this section with an extension of the idea of a group homomorphism. Let D and D' be division rings. A function $\sigma : D \to D'$ is called a *(division ring) monomorphism* if

$$\begin{cases} (\alpha + \beta)\sigma = (\alpha)\sigma + (\beta)\sigma, \text{ and} \\ (\alpha \cdot \beta)\sigma = (\alpha)\sigma \cdot (\beta)\sigma, \ \forall \alpha, \beta \in D, \text{ and } \sigma \neq 0. \end{cases} \qquad (*)$$

If $D = D'$, and if σ maps D *onto* itself, we call σ an *automorphism of D*.

Remarks: There is no need to consider "homomorphisms" of division rings, since it can easily be shown that the "kernel" of any map satisfying (*) must be 0, the zero element of D. We frequently denote the image of α under σ by α^σ. It is always true that $0^\sigma = 0'$ and $1^\sigma = 1'$, for any monomorphism, where 0, 1 and $0'$, $1'$ are defined as usual.

EXAMPLES

(1) In any division ring, the identity map is an automorphism.
(2) The map $\alpha \to \alpha + 0i$ is a monomorphism of $\mathbb{R}$ into $\mathbb{C}$.
(3) "Complex conjugation": $\alpha + i\beta \to \alpha - i\beta$ is an automorphism of $\mathbb{C}$.
(4) The map $\alpha + i\beta \to \alpha$ from $\mathbb{C}$ onto $\mathbb{R}$ is *not* a field monomorphism, although it "preserves addition," and hence is a group homomorphism of the group $\mathbb{C}$ under addition onto the group $\mathbb{R}$ under addition.
(5) The map $\alpha I + \beta i + \gamma j + \delta k \to \alpha I - \beta i - \gamma j - \delta k$ is an automorphism of $\mathcal{Q}$, the real quaternions, defined in Section 1.8.
(6) The map $x + H \to x^p + H$ is an automorphism of Z_p. This is a misleading example. (Explain.)
(7) **(a)** Let $(G, \cdot)$ be a group, $h \in G$. The map $\bar{h} : G \to G$ defined by $(g)\bar{h} = hgh^{-1}$ is an automorphism of G. This automorphism is called an *inner automorphism*.
(b) If D is a division ring and $\alpha \in D$, $\alpha \neq 0$, then the map $\bar{\alpha} : D \to D$ defined by $\beta^{\bar{\alpha}} = \alpha\beta\alpha^{-1}$ is an automorphism of D (called an *inner automorphism*).

Chapter 2
Affine Geometry

SECTION 2.1
THE AFFINE STRUCTURE OF A VECTOR SPACE

Affine geometry refers to the investigation of such concepts as points, lines, planes, incidence, and parallelism. Roughly speaking, the results encountered in high-school geometry that do not require the notion of distance between points constitute the subject matter of affine geometry. As we shall see, some of the results that normally are proved using the distance function can also be treated by present methods.

Our development will continue to be algebraic in nature. For example, "points" will be vectors in ${}_DV$, where D is a division ring. An alternative approach is to begin with a set S of "points," and certain subsets of S called "lines," "planes," and so on, satisfying certain axioms such as:

If P and Q are distinct points in S, then there exists a unique line $L \subseteq S$ such that $P \in L$ and $Q \in L$.

This and other familiar axioms will appear as theorems in our context. Later we shall discuss the reverse procedure, namely, to begin with a set S of "points" and impose sufficiently many axioms to insure that some vector space over some division ring would serve as a *model* for the geometry.

DEFINITION 1 Let ${}_DV$ be a vector space. An *affine subspace* of V is a coset $A = v + U$, where $v \in V$, $U < V$. Also, the empty set $\varnothing$ is considered to be an affine subspace.

If S and T are subsets of V, we put

$$S - T = \{s - t : s \in S,\ t \in T\}.$$

In case T is a singleton set $\{t\}$, we write $S - t$ for $S - \{t\}$.

Proposition 1

Let $A = v + U$ be an affine subspace of V, and let $a_0 \in A$. Then:

(**a**) $U = A - a_0 = A - A$;
(**b**) if $A = v + W$, then $U = W$.

Proof

(**a**) We proved in Chapter 1 that $A = a_0 + U$. Hence $U = A - a_0 \subseteq A - A$. Also if $a - a' \in A - A$, then $a = a_0 + u$, $a' = a_0 + u'$, and $a - a' = u - u' \in U$. That is, $A - A \subseteq U$, giving the desired equalities.
(**b**) From (a) we have $W = A - v = U$.

Corollary

Let A be an affine subspace of V. Then $O \in A \Leftrightarrow A < V$.

Proof

If A is an affine subspace of V, then $O \in A \Rightarrow A \neq \varnothing$. Hence $A = v + U$ for $U < V$. But $U = A - O = A$.

The proposition says that to each nonempty affine subspace A of V there corresponds a *unique* linear subspace, which we denote $L(A)$, in such a way that $A = a + L(A)$, for any $a \in A$.

DEFINITION 2 Let A be an affine subspace of V. The *dimension* of A, denoted by $\dim(A)$, is defined to be $\dim(L(A))$. We also assign $\dim(\emptyset) = -1$. In particular, if $A = v = v + O$, we have $\dim(A) = 0$. Affine subspaces of dimensions 0, 1, and 2 are called *points*, *lines*, and *planes*, respectively. If $\dim(V) = n$, and $\dim(A) = n - 1$, we call A a *hyperplane*. If $\dim(A) = k < n$, we call A a *k-dimensional hyperplane*. The space V itself, when viewed as an affine subspace, will be called an *n-dimensional affine space*.

EXAMPLES

(1) Let $V = \mathbb{R}^n$, $\beta \in \mathbb{R}$, $A = \{(\alpha_1, \cdots, \alpha_n) : \sum_{i=1}^n \alpha_i = \beta\}$. Then A is an affine subspace of V. (See Exercise 1.)

(2) Let V be the polynomials of degree less than or equal to 5 over $\mathbb{R}$, that is, $\mathbb{R}^5[x]$. Put $A = \{p(x) \in V : p(0) = 3,\ p(1) = 4\}$. Then A is a 4-dimensional hyperplane in V. (See Exercise 2.)

The next result gives us a practical test for affine subspaces. Note the resemblance between this criterion and one of the criteria for a (linear) subspace. In $\mathbb{R}^2$ or $\mathbb{R}^3$ the line passing through the points u and v is given by $\{\alpha u + (1 - \alpha)v : \alpha \in \mathbb{R}\}$. An affine subspace of $\mathbb{R}^2$ or $\mathbb{R}^3$ is a subset A with the property that if $u, v \in A$, then the line determined by u and v lies in A; hence:

Proposition 2

Let ${}_DV$ be a vector space, and suppose that $2^{-1} \in D$. A nonempty subset A of V is an affine subspace if and only if $\alpha a_1 + (1 - \alpha)a_2 \in A$, for all $a_1, a_2 \in A$, and all $\alpha \in D$.

Proof

Suppose $\alpha a_1 + (1 - \alpha)a_2 \in A$, for all $\alpha \in D$, $a_1, a_2 \in A$. Let $a_0 \in A$, and put $U = A - a_0$. It suffices to show $U < V$. Let $u_1, u_2 \in U$, with $u_1 = a_1 - a_0$, $u_2 = a_2 - a_0$, for $a_1, a_2 \in A$. For $\alpha \in D$, we have

$$\alpha a_1 + (1 - \alpha)a_0 = \alpha(u_1 + a_0) + (1 - \alpha)a_0 = \alpha u_1 + a_0 \in A,$$

so that $\alpha u_1 \in A - a_0 = U$. Also,

$$\alpha a_1 + (1 - \alpha)a_2 = \alpha(u_1 + a_0) + (1 - \alpha)(u_2 + a_0) = \alpha u_1 + (1 - \alpha)u_2 + a_0 \in A,$$

so that $\alpha u_1 + (1 - \alpha)u_2 \in A - a_0 = U$. In particular $2^{-1}u_1 + 2^{-1}u_2 \in U$, and hence $u_1 + u_2 = 2 \cdot (2^{-1}u_1 + 2^{-1}u_2) \in U$, and thus $U < V$.

Conversely, if A is an affine subspace of V, then $A = a_0 + L(A)$. For $\alpha \in D$, and $a_1, a_2 \in A$, we have $a_1 = a_0 + u_1$, $a_2 = a_0 + u_2$, and with $u_1, u_2 \in L(A)$. Since $L(A)$ is a subspace,

$$\begin{aligned} \alpha a_1 + (1 - \alpha)a_2 &= \alpha(a_0 + u_1) + (1 - \alpha)(a_0 + u_2) \\ &= a_0 + [\alpha u_1 + (1 - \alpha)u_2] \in A. \end{aligned}$$

The hypothesis, $2^{-1} \in D$, deserves some mention. Suppose $D = \mathbb{Z}_2$, the field of two elements $\{0, 1\}$. Then $2^{-1} \notin D$ since $2 = 1 + 1 = 0$ in D. The condition in ${}_DV$, $\alpha a_0 + (1 - \alpha)a_1 \in A$ whenever $\alpha \in D$ and a_0, $a_1 \in A$, is satisfied for any subset A of ${}_DV$, since $\alpha a_0 + (1 - \alpha)a_1$ equals either a_0 (if $\alpha = 1$) or a_1 (if $\alpha = 0$). In view of the fact that we shall be using the criterion in Proposition 2 frequently, *we make the convention that* $2^{-1} \in D$ *throughout the remainder of this chapter*. Several exercises will be provided to illustrate some of the pathologies encountered when $D = \mathbb{Z}_2$.

As was the case with linear subspaces, we can extend the criterion of Proposition 2 to the case where more than two vectors are involved.

Theorem 1

Let A be a subset of ${}_DV$. Then A is an affine subspace of $V \Leftrightarrow \sum_{i=0}^{n} \alpha_i a_i \in A$ whenever $a_0, a_1, \cdots, a_n \in A$, and $\sum \alpha_i = 1$ for $\alpha_0, \alpha_1, \cdots, \alpha_n \in D$.

Proof

Of course if $A = \varnothing$, the condition is fulfilled vacuously, and conversely.

$\boxed{\Leftarrow}$: In particular, when $n = 1$, we have $\alpha_0 a_0 + (1 - \alpha_0)a_1 \in A$, whenever a_0, $a_1 \in A$ and $\alpha_0 \in D$. By Proposition 2, A is an affine subspace.

$\boxed{\Rightarrow}$: Suppose A is an affine subspace. Let $a_0, a_1, \cdots, a_n \in A$, and $\alpha_0, \alpha_1, \cdots, \alpha_n \in D$ with $\sum \alpha_i = 1$. Then, recalling $L(A) = A - a_0$, we have

$$\sum_{i=0}^{n} \alpha_i a_i = \left(\sum_{i=0}^{n} \alpha_i\right) a_0 + \sum_{i=0}^{n} \alpha_i(a_i - a_0) = a_0 + \sum_{i=0}^{n} \alpha_i(a_i - a_0) \in A.$$

In view of Theorem 1, we shall call a linear combination of the form $\alpha_0 a_0 + \alpha_1 a_1 + \cdots + \alpha_n a_n$ an *affine combination of the vectors* $a_0, a_1, \cdots, a_n$ provided $\alpha_0 + \alpha_1 + \cdots + \alpha_n = 1$. Before we exploit this characterization of affine subspaces given in the theorem, we record a fact that helps to distinguish arbitrary affine subspaces from linear subspaces.

As Figure 2.1 indicates, in $\mathbb{R}^2$ any line L not passing through the origin has the property that if $(\alpha_1, \alpha_2) \in L$, then $\beta(\alpha_1, \alpha_2) \notin L$ if $\beta \neq 1$. More generally we have:

Proposition 3

Let A be an affine subspace of ${}_DV$. The following statements are equivalent:

(a) $\exists a_0, a_1, \cdots, a_n \in A$ and $\alpha_0, \alpha_1, \cdots, \alpha_n \in D$ such that $\alpha_0 a_0 + \alpha_1 a_1 + \cdots + \alpha_n a_n \in A$, but $\alpha_0 + \alpha_1 + \cdots + \alpha_n \neq 1$;
(b) $\exists a \in A$ such that $\beta a \in A$ for some $\beta \neq 1$, $\beta \in D$;
(c) $O \in A$;
(d) $A = L(A)$.

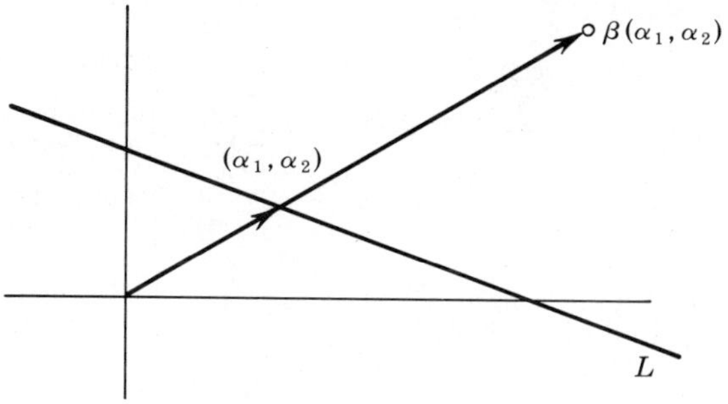

Figure 2.1

Proof

(a) ⇒ (b) Suppose $\sum_{i=0}^{n} \alpha_i a_i \in A$, and $\sum_{i=0}^{n} \alpha_i = \beta \neq 1$. If $\beta = 0$, then $2(\sum_{i=0}^{n} \alpha_i a_i) = 1 \cdot (\sum_{i=0}^{n} \alpha_i a_i) + \sum_{i=0}^{n} \alpha_i a_i \in A$, since $1 + \sum_{i=0}^{n} \alpha_i = 1 + \beta = 1$. Otherwise if $\beta \neq 0$, then $(\beta)^{-1} \sum_{i=0}^{n} \alpha_i a_i = \sum_{i=0}^{n} (\beta)^{-1} \alpha_i a_i \in A$, since $\sum_{i=0}^{n} (\beta)^{-1} \alpha_i = 1$. Thus, in either case, some multiple of $\sum_{i=0}^{n} \alpha_i a_i$ belongs to A.

(b) ⇒ (c) Suppose a, $\beta a \in A$, for $\beta \neq 1$. Put $\alpha = -(1 - \beta)^{-1}\beta$. Then

$$(1 - \alpha) = (1 - \beta)^{-1}(1 - \beta) + (1 - \beta)^{-1}\beta = (1 - \beta)^{-1},$$

and

$$\alpha a + (1 - \alpha)(\beta a) = -(1 - \beta)^{-1}\beta a + (1 - \beta)^{-1}\beta a = O \in A.$$

(c) ⇒ (d) If $O \in A$, then $A = O + L(A) = L(A)$.

(d) ⇒ (a) Since A is linear, $2 \cdot O = O \in A$, and $2 \neq 1$.

For a subset S of a vector space V the following defines the affine analogue of the linear concept, $\langle S \rangle$.

DEFINITION 3 Let $S \subseteq {}_D V$. The *join of* S, denoted $J(S)$, is the set of all affine combinations of finite subsets of S. It will be convenient to denote $J(\{x_1, \cdots, x_n\})$ by $J(x_1, \cdots, x_n)$, for $x_1, \cdots, x_n \in V$. Also we write $J(S_1, \cdots, S_n)$ for $J(S_1 \cup \cdots \cup S_n)$, where $S_1, \cdots, S_n$ are subsets of V.

Proposition 4

Let $S \subseteq {}_D V$. Then $J(S)$ is an affine subspace of V.

Proof

Let $\sum_{i=0}^{m} \alpha_i s_i$ and $\sum_{i=0}^{n} \alpha_i' s_i'$ be elements in $J(S)$. For $\alpha \in D$, we have

$$\alpha \sum_{i=1}^{m} \alpha_i s_i + (1 - \alpha) \sum_{i=0}^{n} \alpha_i' s_i' = \sum_{i=0}^{m} (\alpha\alpha_i) s_i + \sum_{i=0}^{n} ((1 - \alpha)\alpha_i') s_i' \in J(S),$$

since

$$\sum_{i=0}^{m} \alpha\alpha_i + \sum_{i=0}^{n} ((1-\alpha)\alpha_i') = \alpha \sum_{i=0}^{m} \alpha_i + (1-\alpha) \sum_{i=0}^{n} \alpha_i'$$
$$= \alpha \cdot 1 + (1-\alpha) \cdot 1$$
$$= 1.$$

Hence $J(S)$ is an affine subspace.

In $\mathbb{R}^2$ a line L can be represented by $L = \{\alpha u + (1-\alpha)v : \alpha \in \mathbb{R}\}$, where $u, v \in L$, and $u \neq v$. It is easy to show that this representation of each point in L is *unique*, that is,

$$\alpha u + (1-\alpha)v = \beta u + (1-\beta)v \Rightarrow (\alpha - \beta)u = (\alpha - \beta)v \Rightarrow \alpha = \beta.$$

We generalize and extend this fact to:

Proposition 5

Let $x_0, x_1, \cdots, x_n \in {}_D V$. The following statements are equivalent:

(a) $\dim (J(x_0, x_1, \cdots, x_n)) = n$;
(b) $\{x_0 - x_i, x_1 - x_i, \cdots, x_{i-1} - x_i, x_{i+1} - x_i, \cdots, x_n - x_i\}$ is linearly independent for $i = 0, 1, \cdots, n$;
(c) $x_i \notin J(x_0, x_1, \cdots, x_{i-1}, x_{i+1}, \cdots, x_n)$ for $i = 0, 1, \cdots, n$;
(d) $\sum_{i=0}^n \alpha_i x_i = \sum_{i=0}^n \beta_i x_i$, where $\sum \alpha_i = \sum \beta_i = 1$ implies $\alpha_i = \beta_i$ for $i = 0, 1, \cdots, n$.

Proof

Note that

$$J(x_0, x_1, \cdots, x_n) = \left\{ \sum_{j=0}^{n} \alpha_j x_j : \sum \alpha_j = 1 \right\}.$$

Hence

$$L(J(x_0, x_1, \cdots, x_n)) = J(x_0, x_1, \cdots, x_n) - x_i$$
$$= \{\sum \alpha_j x_j - x_i : \sum \alpha_j = 1\}$$
$$= \{\sum \alpha_j x_j - (\sum \alpha_j) x_i : \sum \alpha_j = 1\}$$
$$= \left\{ \sum_{\substack{j=0 \\ j \neq i}}^{n} \alpha_j (x_j - x_i) : \sum_{j \neq i} \alpha_j = 1 - \alpha_i \right\}.$$

We observe that, since α_i is arbitrary, the last expression is actually

$$\langle x_0 - x_i, \cdots, x_{i-1} - x_i, x_{i+1} - x_i, \cdots, x_n - x_i \rangle.$$

It follows that **(a)** $\Leftrightarrow$ **(b)**. The remaining implications are left as an exercise.

Corollary 1

(a) The join of two distinct points is a line.
(b) The join of three noncollinear points is a plane.

Corollary 2

Let $\varnothing \neq A$ be an n-dimensional affine subspace of $_DV$. Then $\exists x_0, x_1, \cdots, x_n \in A$ such that each $a \in A$ can be written uniquely

$$a = \alpha_0 x_0 + \alpha_1 x_1 + \cdots + \alpha_n x_n \qquad (*)$$

as an affine combination of $x_0, x_1, \cdots, x_n$.

Proof

Let $x_0 \in A$. The linear subspace $A - x_0$ has a basis $\{x_1 - x_0, x_2 - x_0, \cdots, x_n - x_0\}$, for some $x_1, \cdots, x_n \in A$.

A set $\{x_0, x_1, \cdots, x_n\}$ of vectors satisfying the conditions of Proposition 5 and Corollary 2 is called *affinely independent*. The coefficients $\{\alpha_0, \alpha_1, \cdots, \alpha_n\}$ in (*) are called the *barycentric coordinates of a relative to* $\{x_0, x_1, \cdots, x_n\}$. This counterpart to the notion of a basis of a linear subspace will be of considerable use to us in the future.

EXERCISES

1. Show that A in Example (1) above is an affine subspace of $\mathbb{R}^n$, and describe $L(A)$.

2. Show that A in Example (2) above is an affine subspace of $\mathbb{R}^5[x]$, and describe $L(A)$. [*Hint*: Find a polynomial $p_1 \in A$ of degree 1, and show that $A = p_1 + L(A)$, where $L(A) = \{p \in V : p(0) = 0 = p(1)\}$].

3. Complete the Proof of Proposition 5.

4. Let $D = \mathbb{Z}_2$. Find a subset of $_DD^2$ that is not a coset. Hence Proposition 2 would fail without the condition $2^{-1} \in D$.

5. Let V, W be vector spaces over D and $f: V \to W$ a homomorphism. If A is an affine subspace of V, show that $(A)f$ is an affine subspace of W. What is $L((A)f)$?

*6. Let $V = \mathbb{R}^{n-1}[x]$, the space of polynomials of degree less than or equal to $n - 1$. Let $\alpha_1, \cdots, \alpha_k$ be distinct real numbers and $\beta_1, \cdots, \beta_k$ be any reals, where $1 \leq k \leq n$. Put $A_i = \{f \in V: f(\alpha_i) = \beta_i\}$, $i = 1, 2, \cdots, k$, and put $A = A_1 \cap \cdots \cap A_k = \{f \in V: f(\alpha_1) = \beta_1, \cdots, f(\alpha_k) = \beta_k\}$.

(a) Show that A_i is an affine subspace of V.
(b) Show that $L(A_i) = \{f \in V: f(\alpha_i) = 0\}$.
(c) Show that A_i is a hyperplane in V. [*Hint*: Observe that $L(A_i) = \{p(x) \cdot (x - \alpha_i): p(x) \in \mathbb{R}^{n-2}[x]\}$, and show that $L(A_i) \approx \mathbb{R}^{n-2}[x]$.]
(d) Show that A is an $(n - k)$-dimensional hyperplane. [*Hint*: Let $p_i(x) = (x - \alpha_1) \cdots (x - \alpha_{i-1})(x - \alpha_{i+1}) \cdots (x - \alpha_k)$. Then $p_i(\alpha_i) \neq 0$ so that there exists $\gamma_i \in \mathbb{R}$ such that $\gamma_i p_i(\alpha_i) = \beta_i$. Show that $\sum_{i=1}^{k} \gamma_i p_i(x) \in A$. This is the existence proof of the Lagrange Interpolation Formula.]
(e) Finally, prove the following fact about arbitrary vector spaces V:

Let $U, W < V$, with W a hyperplane. Then either $U \subseteq W$, or $\dim (U \cap W) = \dim (U) - 1$.

Using this fact and part (c), show that $A = a + \bigcap_{i=1}^{k} L(A_i)$, where a is any element of A. [4.9]

SECTION 2.2
JOINS, INCIDENCE, AND PARALLELISM

In Section 2.1 we defined the join $J(S)$ of a subset S in ${}_DV$. A particular instance of this construction is where $S = A_1 \cup A_2$ for affine subspaces A_1, A_2 of V, where we denoted $J(S)$ by $J(A_1, A_2)$. Just as $J(S)$ is the affine analogue of $\langle S \rangle$ for linear subspaces, so $J(A_1, A_2)$ corresponds to $A_1 + A_2$ in case $A_1, A_2 < V$. This analogy is but part of an extensive comparison between the linear structure and the affine structure of a finite dimensional vector space. We take the following viewpoint.

Let ${}_DV$ be fixed. Consider the collection $\mathscr{L}(V)$ of all linear subspaces of V. We know two binary operations on $\mathscr{L}$, namely intersection and addition. We know further that these operations are linked together via the notion of dimension by the equation

$$\dim (U + W) = \dim (U) + \dim (W) - \dim (U \cap W).$$

(See Chapter 1, Section 1.4, Exercise 7.) Our present goal is to consider the analogous binary operations on the collection $\mathscr{A}(V)$ of all affine subspaces of V, namely, intersection and join. We shall then compare these two structures $\mathscr{L}(V)$ and $\mathscr{A}(V)$ in the exercises as the beginning of a more fundamental investigation—the relationship between the algebraic development and the axiomatic development of geometry.

We begin by showing that intersection is a binary operation on $\mathscr{A}(V)$. As in the linear case we have, more generally:

Proposition 1

Let I be any set, and let $\{A_i : i \in I\}$ be a collection of affine subspaces of V. Then

$$A = \bigcap_{i \in I} A_i$$

is an affine subspace of V. If $A \neq \varnothing$, then

$$A = a + \bigcap_{i \in I} L(A_i)$$

for any $a \in A$.

Proof

If $A = \varnothing$, we are done. Otherwise suppose $a \in A$. Put $L = \bigcap_{i \in I} L(A_i)$. We have already shown that $L < V$. Let $x \in L$. Then $a + x \in A_i$, $\forall i \in I \Rightarrow a + x \in A$, so that $a + L \subseteq A$. Now let $y \in A$. Then $y \in A_i \Rightarrow y - a \in L(A_i)$, $\forall i \in I \Rightarrow y - a \in L$, so that $A - a \subseteq L$, or $A \subseteq a + L$ and thus $A = a + L$.

Proposition 2

Let S be a nonempty subset of V. Then:

(a) $J(S) = s + \langle S - S \rangle = s + \langle S - s \rangle$, for any $s \in S$;
(b) $J(S)$ is the intersection of all affine subspaces of V that contain S; hence $J(S)$ is the smallest affine subspace of V containing S.

Proof

(a) Since $L(J(S)) < V$, and $S \subseteq J(S)$, we have

$$L(J(S)) = \langle J(S) - J(S) \rangle \supseteq \langle S - S \rangle \supseteq \langle S - s \rangle, \forall s \in S. \tag{*}$$

Also $S = s + (S - s) \subseteq s + \langle S - s \rangle$, and the latter is an affine subspace. Hence

$$L(J(S)) = J(S) - s \subseteq \langle S - s \rangle. \tag{**}$$

Combining (*) and (**), we have

$$L(J(S)) \subseteq \langle S - s \rangle \subseteq \langle S - S \rangle \subseteq L(J(S)),$$

yielding(**a**).

(**b**) This follows as an immediate consequence of Theorem 1 and Proposition 4 in Section 2.1.

We showed in Section 2.1 that the join of two distinct points is a line and that the join of three noncollinear points is a plane. The description of $J(A_1, A_2)$, for affine subspaces A_1 and A_2 other than points, is, however, more delicate. For example, the reader's intuition should tell him that the join of two lines in $\mathbb{R}^3$ is either all of $\mathbb{R}^3$ or some plane, depending on whether or not the lines are "skew." Let us proceed to give a more accurate account of the join of subspaces.

Proposition 3

Let A, B be affine subspaces of V, with $a \in A$ and $b \in B$. Then $J(A, B) = a + L(A) + L(B) + \langle b - a \rangle$.

Proof

We know that $C = a + L(A) + L(B) + \langle b - a \rangle$ is an affine subspace. Also, $A = a + L(A) \subseteq C$, and $B = b + L(B) = a + (b - a) + L(B) \subseteq C$. Thus $J(A, B) \subseteq C$. To establish the reverse inclusion we will show $L(J(A, B)) \supseteq L(C)$. Since

$$L(J(A, B)) = J(A, B) - a \supseteq A - a = L(A),$$

and

$$L(J(A, B)) = J(A, B) - b \supseteq B - b = L(B),$$

and

$$L(J(A, B)) = \langle J(A, B) - J(A, B) \rangle \supseteq \langle b - a \rangle,$$

we have

$$L(J(A, B)) \supseteq L(A) + L(B) + \langle b - a \rangle = L(C).$$

Proposition 4

Let A, B be affine subspaces of V, with $a \in A$, and $b \in B$. Then $A \cap B \neq \varnothing \Leftrightarrow (b - a) \in L(A) + L(B)$.

Proof

$A \cap B \neq \varnothing \Leftrightarrow$ there exists $c \in A \cap B \Leftrightarrow c = a + s = b + t$ with $s \in L(A)$, $t \in L(B) \Leftrightarrow b - a = s - t \in L(A) + L(B)$.

Corollary

Let $A = a + L(A)$, $B = b + L(B)$. If $A \cap B \neq \varnothing$, then $J(A, B) = a + L(A) + L(B)$.

Theorem 5

Let V be finite dimensional, and A, B affine subspaces of V. Then:

(a) $\dim J(A, B) = \dim (A) + \dim (B) - \dim (A \cap B)$, if $A \cap B \neq \varnothing$;
(b) $\dim J(A, B) = 1 + \dim (L(A) + L(B))$, if $A \cap B = \varnothing$.

Proof

(a) If $A \cap B \neq \varnothing$, then $\langle b - a \rangle \subseteq L(A) + L(B)$.

$$\begin{aligned}\dim J(A, B) &= \dim (L(A) + L(B) + \langle b - a \rangle) = \dim (L(A) + L(B))\\ &= \dim L(A) + \dim L(B) - \dim (L(A) \cap L(B))\\ &= \dim A + \dim B - \dim (A \cap B),\end{aligned}$$

recalling that $A \cap B \neq \varnothing \Rightarrow L(A \cap B) = L(A) \cap L(B)$, by Proposition 1.

(b) $A \cap B = \varnothing \Rightarrow \langle b - a \rangle \nsubseteq L(A) + L(B)$ so that

$$\begin{aligned}L(J(A, B)) &= L(A) + L(B) + \langle b - a \rangle\\ &= (L(A) + L(B)) + \langle b - a \rangle.\end{aligned}$$

Then

$$\begin{aligned}\dim J(A, B) &= \dim (L(A) + L(B)) + \dim \langle b - a \rangle\\ &= \dim (L(A) + L(B)) + 1.\end{aligned}$$

DEFINITION Let A, B be affine subspaces of V. We say A *is parallel to* B, denoted $A \parallel B$, if either $L(A) \subseteq L(B)$ or $L(B) \subseteq L(A)$.

Remarks: This definition of parallelism agrees with the usual definition for lines in $\mathbb{R}^2$, for planes in $\mathbb{R}^3$, and for a line and a plane in $\mathbb{R}^3$. Although parallelism is an equivalence relation in the set of lines in $\mathbb{R}^2$ as well as the set of planes in $\mathbb{R}^3$, it should be noted that two nonparallel lines L_1 and L_2 in $\mathbb{R}^3$ can both be parallel to a plane M. (See Exercise 1.) From our definition it follows that if $A \subseteq B$, then $A \parallel B$. The next result shows that except for this particular case, parallelism implies disjointness.

Proposition 6

Let A, B be affine subspaces such that $A \nsubseteq B$, $B \nsubseteq A$, and $A \parallel B$. Then $A \cap B = \varnothing$.

Proof

Since $A \parallel B$, we have $L(A) \subseteq L(B)$, say. Then if $c \in A \cap B$, we would have $A = c + L(A) \subseteq c + L(B) = B$, a contradiction. Hence $A \cap B = \varnothing$.

A partial converse to Proposition 6 is available. One might be tempted to expect a better result than the following; if so, see Exercise 2.

Proposition 7

Let A, B be hyperplanes of V (that is, $\dim V = n$, $\dim A = \dim B = n - 1$). Then $A \cap B = \varnothing \Rightarrow A \parallel B$.

Proof

By Theorem 5, $A \cap B = \varnothing \Rightarrow n \geq \dim J(A, B) = \dim (L(A) + L(B)) + 1$, so that

$$\dim (L(A) + L(B)) \leq n - 1.$$

But $L(A) + L(B)$ contains $L(A)$ and $L(B)$, both of dimension $n - 1$; hence $L(A) = L(B) = L(A) + L(B)$ and $A \parallel B$.

EXERCISES

1. (a) Let A, B be affine subspaces of the same dimension. Show that if $A \parallel B$, then $L(A) = L(B)$.
   (b) Let $\dim V = n$ and $0 \leq k \leq n$. Show that parallelism is an equivalence relation on the set of k-dimensional hyperplanes of V.
   (c) Find nonparallel lines L_1, L_2 and a plane M in $\mathbb{R}^3$ such that $L_1 \parallel M$ and $L_2 \parallel M$.

2. In $\mathbb{R}^4$ find two planes that are neither parallel nor have a point in common.

3. Let $A_1, \cdots, A_r$ be subsets of a vector space V. Show that

   $$J(A_1, \cdots, A_r) = J(J(A_1, \cdots, A_k), A_{k+1}, \cdots, A_r), \text{ for } 1 \leq k \leq r.$$

4. Prove the "parallel postulate": Let A be an affine subspace of V. Let $x \in V$, $x \notin A$. Then there exists a unique affine subspace B in V such that $x \in B$, $B \parallel A$, and $\dim B = \dim A$.

5. (a) Show that the join of two affine subspaces is not necessarily the union of all lines joining points in one to points in the other by considering a line and a point not on it in $\mathbb{R}^2$.
   (b) Show that if A, B are affine subspaces and $A \cap B \neq \varnothing$, then $J(A, B)$ is the union of all lines joining points of A to points of B.

DEFINITION A *lattice* is a set $\mathscr{L}$ together with two associative commutative binary operations, denoted "$\cap$" and "$\cup$" such that $x \cup x = x \cap x = x$, $\forall x \in \mathscr{L}$, and $x \cap (x \cup y) = x \cup (x \cap y) = x$, $\forall x, y \in \mathscr{L}$.

Note that if $\mathscr{L}$ is a lattice then we can define an "order relation" between some pairs of elements of $\mathscr{L}$ in the following way:

$$x \subseteq y \text{ if } x \cap y = x.$$

The reader may check that this relation satisfies the usual axioms for a *partial ordering*:

$$x \subseteq x,$$
$$x \subseteq y \text{ and } y \subseteq x \Rightarrow x = y,$$
$$x \subseteq y \text{ and } y \subseteq z \Rightarrow x \subseteq z.$$

Moreover, we could define upper and lower bounds on subsets of $\mathscr{L}$ and then least upper bounds (l.u.b.) and greatest lower bounds (g.l.b.), and show finally that the l.u.b. (respectively g.l.b.) of x, y is $x \cup y$ (respectively $x \cap y$).

*6. Let ${}_DV$ be a finite dimensional vector space. Let $\mathscr{L}(V)$ be the totality of (linear) subspaces of V, and let $\mathscr{A}(V)$ be the totality of affine subspaces of V.

(a) Show that $\mathscr{L}(V)$ is a lattice where "$\cap$" and "$\cup$" are respectively set intersection and addition. Also describe the ordering that results.
(b) Show that $\mathscr{A}(V)$ is a lattice where "$\cap$" and "$\cup$" are respectively set intersection and join. Describe the ordering "$\subseteq$" that results. [4.4]

DEFINITION A lattice $\mathscr{L}$ is said to be *distributive* if

$$x \cap (y \cup z) = (x \cap y) \cup (x \cap z), \forall x, y, z \in \mathscr{L}.$$

A lattice is said to be *modular* if

$$y \subseteq x \Rightarrow x \cap (y \cup z) = y \cup (x \cap z).$$

*7. (a) Show that neither $\mathscr{L}(V)$ nor $\mathscr{A}(V)$ is distributive in general.
(b) Show that $\mathscr{L}(V)$ is modular, but $\mathscr{A}(V)$ is not. [4.4]

DEFINITION Let $\mathscr{L}$ be a lattice. A subset $\mathscr{L}'$ of $\mathscr{L}$ is called a *chain* if

$$x_1, x_2 \in \mathscr{L}' \Rightarrow x_1 \subseteq x_2 \text{ or } x_2 \subseteq x_1.$$

We refer to a (finite) sequence $\{x_0, \cdots, x_k\}$ as a *proper chain* if

$$x_i \subseteq x_{i+1}, \text{ but } x_i \neq x_{i+1}, \text{ for } i = 0, 1, \cdots, k.$$

We say that a proper chain $\{x_0, \cdots, x_k\}$ has *length* k, and we define the *height* of an element $x \in \mathscr{L}$ to be the maximum length of a proper chain $\{x_0, \cdots, x_k = x\}$ having x for its greatest element.

*8. Show that each element in $\mathscr{L}(V)$ (respectively $\mathscr{A}(V)$) has a (finite) height equal to its dimension (respectively, its dimension minus 1). [4.4]

SECTION 2.3
SOME GEOMETRICAL THEOREMS

This section is, in a sense, a major departure from the development of the rest of the chapter. Elsewhere our primary objective is the development of those aspects of the affine geometry of a vector space that are needed in the later axiomatic formulation of geometry. In this section we will show you that geometric theorems, of the type a high-school geometry course leads you to expect, can be formulated and proved in our affine context.

An immediate difficulty seems to be posed by the fact that such a large number of the theorems of high-school geometry involve distance and we have no notion of distance available. However, careful examination shows that in many cases it is not distance but the ratio of distances that is involved, and this, as we now see, can be translated into the affine context.

Proposition 1

Let ${}_DV$ be a vector space, $p, q \in V$ distinct points, $x \in J(p, q)$, $x \neq q$. Then there exists a unique $\alpha \in D$, $\alpha \neq 1$ such that $p - x = \alpha(q - x)$. Conversely, given $p, q \in V$, $\alpha \in D$, $\alpha \neq 1$, there exists a unique $x \in J(p, q)$, $x \neq q$ such that $p - x = \alpha(q - x)$.

Proof

If $p = x$ then $p - x = 0 \cdot (q - x)$. If p, q, x are distinct then $J(p, x) = J(q, x) \Rightarrow L(J(p, x)) = L(J(q, x)) \Rightarrow \langle p - x \rangle = \langle q - x \rangle$, so there exists $\alpha \in D$ such that $p - x = \alpha(q - x)$. For the converse let $x = (1 - \alpha)^{-1}p - (1 - \alpha)^{-1}\alpha q$. $(1 - \alpha)^{-1} - (1 - \alpha)^{-1}\alpha = 1$, so $x \in J(p, q)$ and easy calculation shows that $p - x = \alpha(q - x)$.

DEFINITION 1 Let $p, q, x \in {}_DV$, $p \neq q$, $x \neq q$, $x \in J(p, q)$. The unique $\alpha \in D$ such that $p - x = \alpha(q - x)$ is denoted by $\alpha = \alpha(p, q{:}x)$ and is called the *ratio* of the three collinear points p, q, x.

It should be noted that Proposition 1 not only establishes the existence of $\alpha(p, q{:}x)$, but also relates it to barycentric coordinates, that is,

$$\alpha = \alpha(p, q{:}x) \Leftrightarrow x = (1 - \alpha)^{-1}p - (1 - \alpha)^{-1}\alpha q.$$

If we interpret scalar multiplication in $\mathbb{R}^2$ as suggested in the intuitive discussion of Chapter 1, Section 1.1, then $\alpha = \alpha(p, q{:}x)$ means, for collinear $p, q, x \in \mathbb{R}^2$, that $p - x$ is α times as long as $q - x$. This interpretation allows us to adapt a number of results from $\mathbb{R}^2$ to ${}_DV$.

DEFINITION 2 Let ${}_DV$ be a vector space. A *triangle* is a set $\Delta = (p_1, p_2, p_3)$ of three affine independent points of V.

It is important to note that we have defined a triangle to be its "vertices." You are accustomed to talking about the sides of a triangle, but these are line segments and we cannot define "segment" unless D is ordered. In sketches we will customarily draw in the sides for their intuitive value, but you should not assume that because we draw them they can be formulated in our context.

Similar triangles are characterized in elementary geometry in terms of the ratio of length of corresponding sides, so we should hope to be able to generalize some theorems about similar triangles to the affine context. The following is an example of such a generalization.

Proposition 2

Let ${}_DV$ be a vector space, $\Delta = (p_1, p_2, p_3)$ a triangle, $x \in J(p_1, p_2)$, $y \in J(p_1, p_3)$, $x \neq p_1, p_2$, $y \neq p_1, p_3$. Then $J(x, y) \parallel J(p_2, p_3) \Leftrightarrow \alpha(p_1, p_2{:}x) = \alpha(p_1, p_3{:}y)$. (This is simply the Euclidean theorem that (p_1, x, y) similar to $(p_1, p_2, p_3) \Leftrightarrow$ the line from x to y is parallel to the line from p_2 to p_3. See Figure 2.2.)

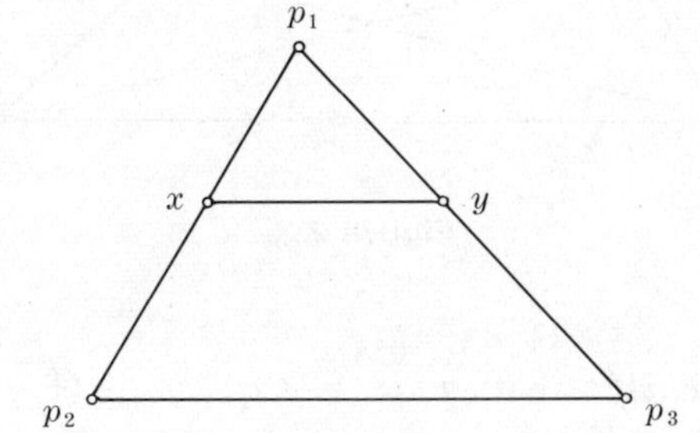

Figure 2.2

Proof

Let $\alpha_1 = \alpha(p_1, p_2 : x)$, $\alpha_2 = \alpha(p_1, p_3 : y)$. Then

$$x = (1-\alpha_1)^{-1}p_1 - (1-\alpha_1)^{-1}\alpha_1 p_2 \quad \text{and} \quad y = (1-\alpha_2)^{-1}p_1 - (1-\alpha_2)^{-1}\alpha_2 p_3.$$

$$\begin{aligned} J(x, y) \parallel J(p_2, p_3) &\Leftrightarrow \text{there is a } \beta \in D \text{ such that } x - y = \beta(p_2 - p_3) \\ &\Leftrightarrow x = y + \beta p_2 - \beta p_3 \\ &\Leftrightarrow x = (1-\alpha_2)^{-1}p_1 - (1-\alpha_2)^{-1}\alpha_2 p_3 + \beta p_2 - \beta p_3 \\ &\Leftrightarrow \text{(using the uniqueness of barycentric coordinates)} \\ &\quad\; (1-\alpha_2)^{-1} = (1-\alpha_1)^{-1}, \qquad \beta = -(1-\alpha_1)^{-1}\alpha_1 \\ &\Leftrightarrow \alpha_1 = \alpha_2. \end{aligned}$$

Since we have no notion of distance available it often comes as something of a surprise that midpoints can easily be defined; if $p, q \in {}_DV$ the *midpoint of p and q* is the $x \in J(p, q)$ such that $x = 2^{-1}p + 2^{-1}q$, which is the x such that $\alpha(p, q : x) = -1$. One can now define the medians of a triangle and hope to prove that they intersect at a point. However, in classical geometry there is a stronger result, known as Ceva's theorem, which we shall generalize and prove.

Theorem 1 (Ceva)

Let $p_1, p_2, p_3, r_1, r_2, r_3$ be six distinct points of ${}_DV$, (p_1, p_2, p_3) a triangle, $r_1 \in J(p_2, p_3)$, $r_2 \in J(p_1, p_3)$, $r_3 \in J(p_1, p_2)$, $\alpha_1 = \alpha(p_2, p_3 : r_1)$, $\alpha_2 = \alpha(p_3, p_1 : r_2)$, $\alpha_3 = \alpha(p_1, p_2 : r_3)$. (See Figure 2.3.) Then $\alpha_1\alpha_2\alpha_3 = -1 \Leftrightarrow$ either $J(p_1, r_1) \parallel J(p_2, r_2) \parallel J(p_3, r_3)$, or the three lines intersect at a point.

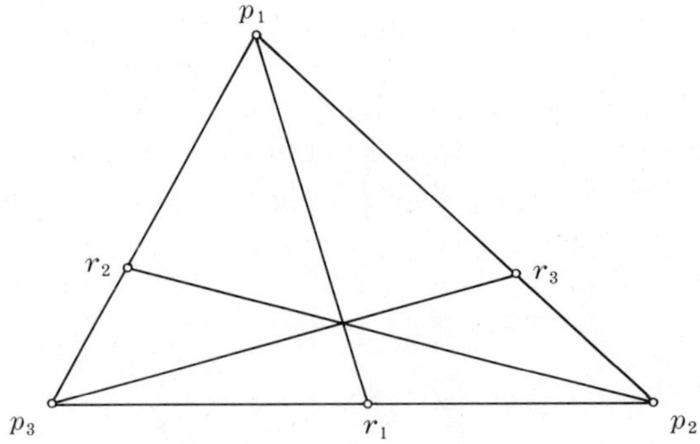

Figure 2.3

Proof

We have immediately that $r_1 = (1-\alpha_1)^{-1}p_2 - (1-\alpha_1)^{-1}\alpha_1 p_3$, $r_2 = (1-\alpha_2)^{-1}p_3 - (1-\alpha_2)^{-1}\alpha_2 p_1$, $r_3 = (1-\alpha_3)^{-1}p_1 - (1-\alpha_3)^{-1}\alpha_3 p_2$. Note also that all points and lines under discussion are in the plane $J(p_1, p_2, p_3)$

and every point $x \in J(p_1, p_2, p_3)$ has a unique representation $x = \gamma_1 p_1 + \gamma_2 p_2 + \gamma_3 p_3$ with $\gamma_1, \gamma_2, \gamma_3 \in D$ and $\gamma_1 + \gamma_2 + \gamma_3 = 1$.

$$\begin{aligned} J(p_1, r_1) &= \{(1-\lambda)p_1 + \lambda r_1 : \lambda \in D\} \\ &= \{(1-\lambda)p_1 + \lambda(1-\alpha_1)^{-1}p_2 - \lambda(1-\alpha_1)^{-1}\alpha_1 p_3 : \lambda \in D\}, \end{aligned}$$

so

$$\begin{aligned} x &= \gamma_1 p_1 + \gamma_2 p_2 + \gamma_3 p_3 \\ &= \gamma_1 p_1 + \gamma_2(1-\alpha_1)(1-\alpha_1)^{-1}p_2 + \gamma_3 \alpha_1{}^{-1}(1-\alpha_1)(1-\alpha_1)^{-1}\alpha_1 p_3 \end{aligned}$$

and $x \in J(p_1, r_1) \Leftrightarrow \gamma_2 = -\gamma_3 \alpha_1{}^{-1}$ using the uniqueness of barycentric coordinates. Similar computations yield

$$x \in J(p_2, r_2) \Leftrightarrow -\gamma_1 \alpha_2{}^{-1} = \gamma_3$$

and

$$x \in J(p_3, r_3) \Leftrightarrow \gamma_1 = -\gamma_2 \alpha_3{}^{-1}.$$

We then have $x \in J(p_1, r_1) \cap J(p_2, r_2) \Leftrightarrow$

$$x = \gamma_1 p_1 + \gamma_2 p_2 + \gamma_3 p_3 = -\gamma_3 \alpha_2 p_1 - \gamma_3 \alpha_1{}^{-1} p_2 + \gamma_3 p_3$$

and

$$1 = -\gamma_3 \alpha_2 - \gamma_3 \alpha_1{}^{-1} + \gamma_3 = \gamma_3(1 - \alpha_1{}^{-1} - \alpha_2);$$

so

$$J(p_1, r_1) \parallel J(p_2, r_2) \Leftrightarrow 1 - \alpha_1{}^{-1} - \alpha_2 = 0.$$

(We are in a plane, so two lines are parallel $\Leftrightarrow$ they do not intersect.) Similarly,

$$J(p_2, r_2) \parallel J(p_3, r_3) \Leftrightarrow 1 - \alpha_3 - \alpha_2{}^{-1} = 0.$$

Case 1

$J(p_1, r_1) \parallel J(p_2, r_2)$. Then

$$\begin{aligned} J(p_2, r_2) \parallel J(p_3, r_3) &\Leftrightarrow 1 - \alpha_3 - \alpha_2{}^{-1} = 0 \\ &\Leftrightarrow \alpha_2 - \alpha_2\alpha_3 - 1 = 0 \Leftrightarrow \alpha_1\alpha_2 - \alpha_1\alpha_2\alpha_3 = \alpha_1 \\ &\Leftrightarrow \alpha_1\alpha_2\alpha_3 = \alpha_1(\alpha_2 - 1). \end{aligned}$$

But $1 - \alpha_1{}^{-1} - \alpha_2 = 0$ so $\alpha_1{}^{-1} = \alpha_1\alpha_2 \Rightarrow -1 = \alpha_1(\alpha_2 - 1)$ so $J(p_2, r_2) \parallel J(p_3, r_3) \Leftrightarrow \alpha_1\alpha_2\alpha_3 = -1$.

Case 2

$J(p_1, r_1) \cap J(p_2, r_1) = x$. Then

$$x = -(1 - \alpha_1{}^{-1} - \alpha_2)^{-1}\alpha_2 p_1 - (1 - \alpha_1{}^{-1} - \alpha_2)^{-1}\alpha_1{}^{-1} p_2 + (1 - \alpha_1{}^{-1} - \alpha_2) p_3$$

by our previous computations. The proof is complete if we show $x \in J(p_3, r_3) \Leftrightarrow \alpha_1\alpha_2\alpha_3 = -1$. But

$$\begin{aligned} x \in J(p_3, r_3) &\Leftrightarrow -(1 - \alpha_1{}^{-1} - \alpha_2)^{-1}\alpha_2 = (1 - \alpha_1{}^{-1} - \alpha_2)\alpha_1{}^{-1}\alpha_3{}^{-1} \\ &\Leftrightarrow -\alpha_2 = \alpha_1{}^{-1}\alpha_3{}^{-1} \\ &\Leftrightarrow \alpha_1\alpha_2\alpha_3 = -1. \end{aligned}$$

Corollary

The medians of a triangle intersect at a point if $3^{-1} \in D$.

Proof

This is simply Ceva's theorem with $\alpha_1 = \alpha_2 = \alpha_3 = -1$. The reader should carefully examine the proof to see the relevance of the condition that $3^{-1} \in D$.

EXERCISES

1. Let ${}_DV$ be a vector space, M_1, M_2 parallel lines in V, and L_1, L_2, L_3 three lines of V intersecting M_1 and M_2. Let $L_i \cap M_1 = p_i$, $L_i \cap M_2 = q_i$, $i = 1, 2, 3$, p_i, q_i all distinct. Show that $L_1 \parallel L_2 \parallel L_3 \Leftrightarrow \alpha(p_1, p_2 : p_3) = \alpha(q_1, q_2 : q_3)$.

2. Let ${}_DV$ be a vector space. A *quadrilateral* is defined to be a set of four points (p_1, p_2, p_3, p_4) any three of which are affine independent. The quadrilateral (p_1, p_2, p_3, p_4) is called a *parallelogram* if $J(p_1, p_2) \parallel J(p_3, p_4)$ and $J(p_1, p_4) \parallel J(p_2, p_3)$. Let r_1 be the midpoint of p_1, p_2; r_2 the midpoint of p_2, p_3; r_3 the midpoint of p_3, p_4; and r_4 the midpoint of p_4, p_1. Show that if (p_1, p_2, p_3, p_4) is a quadrilateral then (r_1, r_2, r_3, r_4) is a parallelogram. Note that p_1, p_2, p_3, p_4 do not have to lie in a plane. What about r_1, r_2, r_3, r_4? If (p_1, p_2, p_3, p_4) is a parallelogram show that the *diagonals* $J(p_1, p_3)$, $J(p_2, p_4)$ intersect at their midpoints, that is, the midpoints of p_1 and p_3 and also of p_2 and p_4.

*3. Given three noncollinear points p_1, p_2, p_3:

   (a) show that there exist p_4, p_4' such that (p_1, p_2, p_3, p_4) and (p_1, p_2, p_4', p_3) are parallelograms;
   (b) show p_4, p_4' are unique. [2.5]

4. With p_1, p_2, p_3, r_1, r_2, r_3, $\alpha_1, \alpha_2, \alpha_3$ as in Ceva's theorem, show that $r_1 \in J(r_2, r_3) \Leftrightarrow \alpha_1 \alpha_2 \alpha_3 = 1$. This is known classically as Menelaus' theorem.

5. If p, q, x are distinct but collinear points, find the relations between $\alpha(p, q : x)$, $\alpha(q, p : x)$, $\alpha(x, p : q)$, $\alpha(p, x : q)$, $\alpha(q, x : p)$, and $\alpha(x, q : p)$.

6. Show by example in $(\mathbb{Z}_3)^2$ that the medians of a triangle need not intersect in a point.

SECTION 2.4
THE DESARGUES THEOREM

The present stage of our development of the affine structure of a vector space allows us to prove a version of the Desargues theorem. The result enjoys full generality, as we shall later see, in the context of projective geometry.

Although we are working over an arbitrary division ring D, we shall use a picture to guide our remarks. It should be noted, however, that any conclusions we draw depend solely on the algebraic facts at our disposal. (For example: The join of two intersecting lines is a plane.)

We say two triangles $\Delta_1 = (p_1, p_2, p_3)$ and $\Delta_2 = (q_1, q_2, q_3)$ are *affine point perspective* if for some pairing, say $(p_1, q_1), (p_2, q_2), (p_3, q_3)$ we have $J(p_1, q_1) \cap J(p_2, q_2) \cap J(p_3, q_3) = p$, a single point.

Theorem 1 (*Desargues*)

Let Δ_1, Δ_2 be perspective as above. Put $L_1 = J(p_2, p_3)$, $L_2 = J(p_1, p_3)$, $L_3 = J(p_1, p_2)$, $L_1' = J(q_2, q_3)$, $L_2' = J(q_1, q_3)$, $L_3' = J(q_1, q_2)$. If for each i, $L_i \cap L_i' = s_i$ (one point), then s_1, s_2, s_3 are collinear. (When this conclusion holds, we say that Δ_1 and Δ_2 are *affine line perspective*.)

Proof

By hypothesis, there exist $\alpha_1, \alpha_2, \alpha_3 \in D$ such that

$$p = \alpha_1 p_1 + (1 - \alpha_1)q_1 = \alpha_2 p_2 + (1 - \alpha_2)q_2 = \alpha_3 p_3 + (1 - \alpha_3)q_3 .$$

(See Figure 2.4.) We first observe that no two of the α_i are equal. For if $\alpha_1 = \alpha_2$, then $\alpha_1(p_1 - p_2) = -(1 - \alpha_1)(q_1 - q_2)$, giving $L_3 \parallel L_3'$, a contradiction. We now locate s_1 in terms of the above data. Since $s_1 \in L_1 \cap L_1'$, we must find $\beta, \gamma \in D$ such that $s_1 = \beta p_2 + (1 - \beta)p_3 = \gamma q_2 + (1 - \gamma)q_3$. But $\alpha_2 p_2 - \alpha_3 p_3 = -(1 - \alpha_2)q_2 + (1 - \alpha_3)q_3$, and multiplying by $(\alpha_2 - \alpha_3)^{-1}$ we find that the coefficients add to 1, so that

$$\begin{aligned} s_1 &= (\alpha_2 - \alpha_3)^{-1}\alpha_2 p_2 - (\alpha_2 - \alpha_3)^{-1}a_3 p_3 \\ &= -(\alpha_2 - \alpha_3)^{-1}(1 - \alpha_2)q_2 + (\alpha_2 - \alpha_3)^{-1}(1 - \alpha_3)q_3 . \end{aligned}$$

Similarly,

$$\begin{aligned} s_2 &= (\alpha_3 - \alpha_1)^{-1}\alpha_3 p_3 - (\alpha_3 - \alpha_1)^{-1}\alpha_1 p_1 \\ &= -(\alpha_3 - \alpha_1)^{-1}(1 - \alpha_3)q_3 + (\alpha_3 - \alpha_1)^{-1}(1 - \alpha_1)q_1, \end{aligned}$$

and

$$\begin{aligned} s_3 &= (\alpha_1 - \alpha_2)^{-1}\alpha_1 p_1 - (\alpha_1 - \alpha_2)^{-1}\alpha_2 p_2 \\ &= -(\alpha_1 - \alpha_2)^{-1}(1 - \alpha_1)q_1 + (\alpha_1 - \alpha_2)^{-1}(1 - \alpha_2)q_2 . \end{aligned}$$

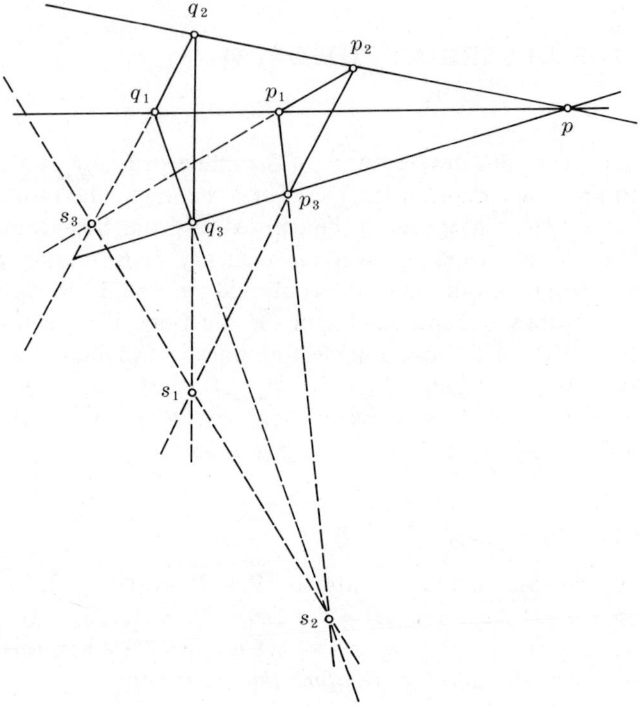

Figure 2.4

Now we wish to show that $s_3 = \delta s_1 + (1 - \delta)s_2$ for some $\delta \in D$. Note that

$$(\alpha_2 - \alpha_3)s_1 + (\alpha_3 - \alpha_1)s_2 + (\alpha_1 - \alpha_2)s_3 = 0,$$

so that

$$s_3 = -(\alpha_1 - \alpha_2)^{-1}(\alpha_2 - \alpha_3)s_1 - (\alpha_1 - \alpha_2)^{-1}(\alpha_3 - \alpha_1)s_2,$$

and the coefficients add to 1.

Remarks. In the statement of the theorem, assumptions are made about the existence of s_1, s_2, and s_3. We can extend the theorem by considering what happens when some of these points fail to exist. For example, if $L_1 \parallel L_1'$ (see Figure 2.5), then s_1 does not exist. The appropriate conclusion is that $J(s_2, s_3) \parallel L_1$ (see Exercise 2). In case two of the pairs of sides are parallel, say $L_1 \parallel L_1'$ and $L_2 \parallel L_2'$, then the conclusion is that $L_3 \parallel L_3'$ also. A reasonably elegant treatment of all these cases can be made by introducing "points at infinity." Briefly, to each class of parallel lines there corresponds one point at infinity which "lies on" each line in that class. Each plane, then, contains one "line at infinity" consisting of all the points at infinity corresponding to

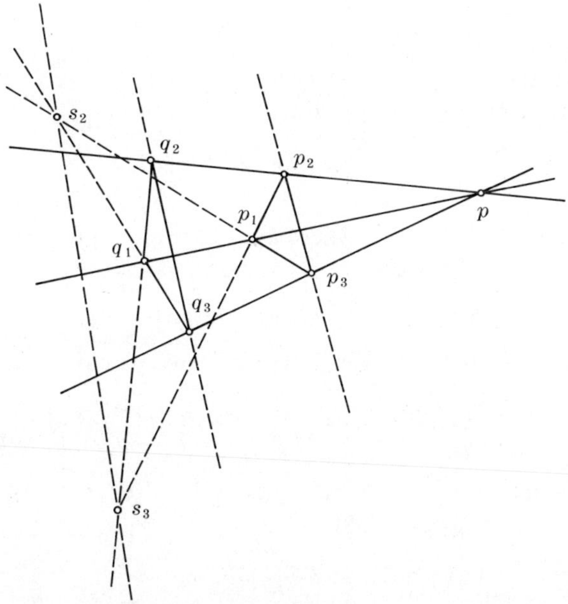

Figure 2.5

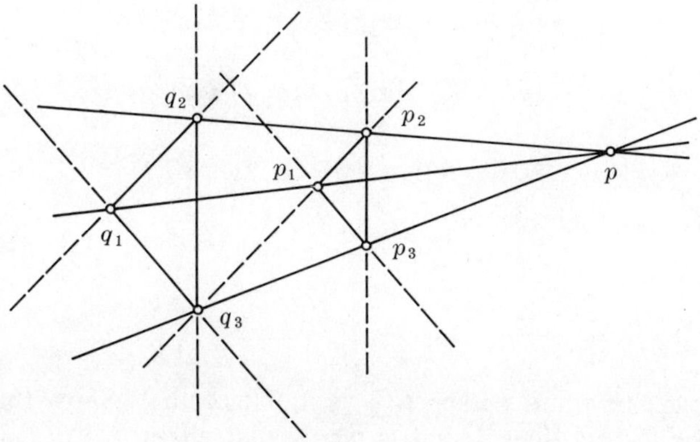

Figure 2.6

parallel classes of lines in that plane. Two planes share a line at infinity if they are parallel. With this terminology, Figure 2.5 represents the case where s_1 is the (unique) point at infinity on the line $J(s_2, s_3)$, and Figure 2.6 represents the case where s_1, s_2, and s_3 are three points on the same line at infinity. A full-scale discussion of these concepts is tantamount to doing projective geometry, a topic we shall develop later.

In connection with this idea of points at infinity, we mention another form of Desargues' theorem. We say that two triangles (p_1, p_2, p_3) and (q_1, q_2, q_3) are *point perspective at infinity* if $J(p_1, q_1) \parallel J(p_2, q_2) \parallel J(p_3, q_3)$. We assert:

Theorem 2

Let (p_1, p_2, p_3) and (q_1, q_2, q_3) be triangles which are point perspective at infinity. Then they are line perspective.

Proof

We sketch the proof of the case where $L_i \cap L_i' = s_i$, using the notation from the earlier version of the theorem. By hypothesis, $\exists \alpha_1, \alpha_2, \alpha_3 \in D$ such that

$$\alpha_1(p_1 - q_1) = \alpha_2(p_2 - q_2) = \alpha_3(p_3 - q_3).$$

We note that $\alpha_i \neq \alpha_j$ for $i \neq j$ since otherwise we would have, for example,

$$\alpha_1(p_1 - q_1) = \alpha_1(p_2 - q_2) \Rightarrow p_1 - q_1 = p_2 - q_2 \Rightarrow p_1 - p_2 = q_1 - q_2 \Rightarrow L_1 \cap L_1' = \varnothing.$$

Once again we locate s_1. Since $s_1 \in J(p_2, p_3) \cap J(q_2, q_3)$, we observe that $\alpha_2 p_2 - \alpha_3 p_3 = \alpha_2 q_2 - \alpha_3 q_3$, yielding

$$s_1 = (\alpha_2 - \alpha_3)^{-1}\alpha_2 p_2 - (\alpha_2 - \alpha_3)^{-1}\alpha_3 p_3 = (\alpha_2 - \alpha_3)^{-1}\alpha_2 q_2 - (\alpha_2 - \alpha_3)^{-1}\alpha_3 p_3.$$

Similarly,

$$s_2 = (\alpha_1 - \alpha_3)^{-1}\alpha_1 p_1 - (\alpha_1 - \alpha_3)^{-1}\alpha_3 q_3,$$

and

$$s_3 = (\alpha_1 - \alpha_2)^{-1}\alpha_1 p_1 - (\alpha_1 - \alpha_2)^{-1}\alpha_2 p_2.$$

Thus

$$(\alpha_2 - \alpha_3)s_1 - (\alpha_1 - \alpha_3)s_2 + (\alpha_1 - \alpha_2)s_3 = O,$$

and $s_1 \in J(s_2, s_3)$. (See Exercise 4.)

EXERCISES

1. Let Δ_1, Δ_2 be point perspective as in Theorem 1. Show that L_i and L_i' cannot be "skew lines," that is, show that either $L_i \cap L_i' =$ a point, or $L_i \parallel L_i'$. Thus the next two exercises, together with Theorem 1, cover all possible situations.

2. In Theorem 1, replace the hypothesis $L_1 \cap L_1' = s_1$ with $L_1 \parallel L_1'$. Prove that $J(s_2, s_3) \parallel L_1$. [*Hint*: Show that $\alpha_2 = \alpha_3$.]

3. In Theorem 1, replace $L_1 \cap L_1' = s_1$ and $L_2 \cap L_2' = s_2$ with $L_1 \parallel L_1'$ and $L_2 \parallel L_2'$. Prove that $L_3 \parallel L_3'$.

4. (a) Fill in the details for the proof of Theorem 2.
   (b) State and prove the analogues of Exercises 1, 2, and 3 for Theorem 2.

SECTION 2.5

THE SIGNIFICANCE OF THE DESARGUES THEOREM

By using the different forms of the Desargues theorem, we can construct various maps of an affine plane onto itself sending points to points and lines to lines. Throughout this discussion we use many familiar-sounding facts that have already been verified, such as the parallel postulate. We also make free use of pictures to illustrate these constructions, but as usual you should note that the constructions depend only upon the stated geometric facts.

Construction 1

Let x, x' be distinct points in A, where A is any affine plane in any space ${}_DV$. Let $y \in A$, $y \notin J(x, x')$. Let y' be the unique point of intersection of the line through x' parallel to $J(x, y)$ and the line through y parallel to $J(x, x')$. In short, y' is the unique point such that (x, x', y', y) is a parallelogram (Section 2.3, Exercise 3). For the remainder of this construction, consult Figure 2.7. Let $z \in A$, $z \notin J(x, x')$, $z \notin J(y, y')$. Such an element z exists since

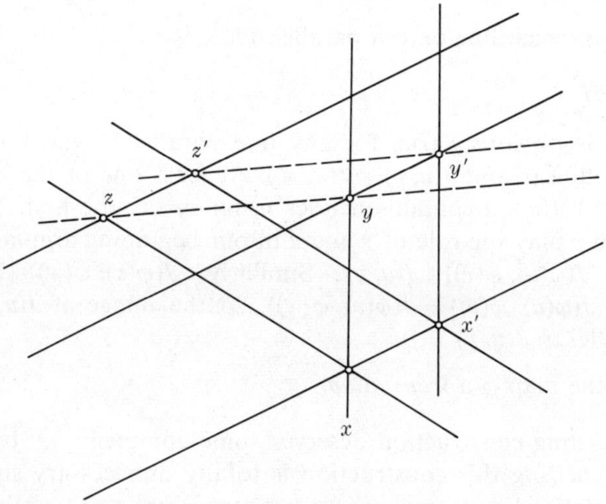

Figure 2.7

we are assuming that $D \neq \mathbb{Z}_2$, which guarantees that each line contains at least three points; thus any line not parallel to $J(x, x')$ contains a point not on $J(x, x')$ and not on $J(y, y')$. Define z' to be the point such that (x, x', z', z) is a parallelogram.

We now have $J(x, x') \parallel J(y, y') \parallel J(z, z')$, as well as $J(x, z) \parallel J(x', z')$ and $J(x, y) \parallel J(x', y')$. But these parallelism relations are precisely the hypotheses of a version of Desargues' theorem (triangles (x, y, z) and (x', y', z') perspective at infinity, and two pairs of corresponding sides parallel), whose conclusion is $J(y, z) \parallel J(y', z')$. Thus we have that (y, y', z', z) is also a parallelogram.

Now let $u \in A$. If $u \notin J(x, x')$, we define $\varphi(u)$ to be the unique point such that $(x, x', \varphi(u), u)$ is a parallelogram. If $u \notin J(y, y')$, we define $\varphi(u)$ to be the unique point such that $(y, y', \varphi(u), u)$ is a parallelogram. If $u \notin J(z, z')$ we define $\varphi(u)$ to be the unique point so that $(z, z', \varphi(u), u)$ is a parallelogram. The above argument shows that any two of these definitions that are applicable do give the same result $\varphi(u)$, that is, the map φ is well defined on all points of A. Moreover $\varphi(x) = x'$, $\varphi(y) = y'$, and $\varphi(z) = z'$.

The map φ has the following properties:

(1) φ is 1-1 and onto.

Proof

Let $u' \in A$. Say $u' \notin J(x, x')$. (Otherwise use $J(y, y')$.) By Section 2.3, Exercise 3, there is a unique point $u \in A$ such that (x, x', u', u) is a parallelogram. Clearly $\varphi(u) = u'$. Thus φ maps A *onto* A. Moreover, this uniqueness of u forces φ to be 1-1.

(2) φ maps each line onto a parallel line.

Proof

This is obviously true for any line parallel to $J(x, x')$. Let u, v, w be collinear and $J(u, v) \nparallel J(x, x')$. At least one of the lines $J(x, x')$, $J(y, y')$, $J(z, z')$ contains neither u nor v, say the first. Then, letting u and v play the role of y and z in our beginning argument, we have that $J(\varphi(u), \varphi(v)) \parallel J(u, v)$. Similarly, $J(\varphi(v), \varphi(w)) \parallel J(u, v)$. But then $J(\varphi(u), \varphi(v)) \doteq J(\varphi(v), \varphi(w))$, so the image of $J(u, v)$ is a line parallel to $J(u, v)$.

We call the map φ a *translation*.

The preceding construction deserves some comment. To begin with, in our algebraic setting this construction is totally unnecessary since the existence of translations and their salient properties are quite easily established (see Exercise 1). However, it should be clear that a rather modest set of axioms is required to perform this construction in a nonalgebraic context (roughly, a sufficient number of points, the parallel postulate, and Desargues' theorem). This observation will be of use to us in Chapter 4. It is also relatively simple now to develop analogous maps in a natural way from other versions of Desargues' theorem.

Construction 2

Let A be an affine plane. Let p, x, x' be distinct collinear points in A. For $y \in A$, $y \notin J(p, x)$ let $y' = J(p, y) \cap L$, where L is the line through x' parallel to $J(x, y)$ (see Figure 2.8). Let $z \in A$, $z \notin J(p, x)$, $z \notin J(p, y)$. Define $z' =$

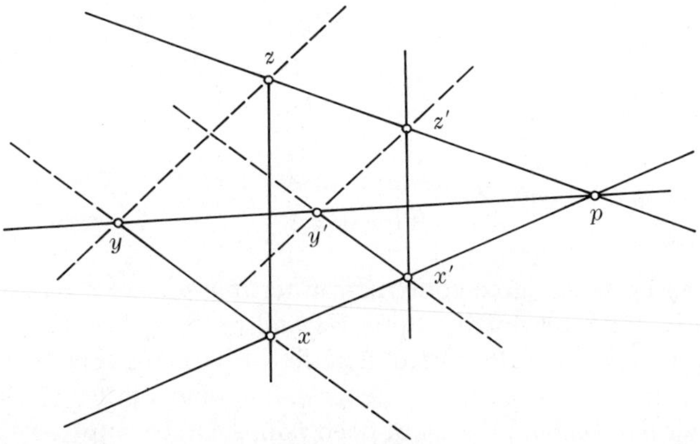

Figure 2.8

$J(p, z) \cap L'$, where L' is the line through x' parallel to $J(x, z)$. The triangles (x, y, z) and (x', y', z') are thus affine perspective at p, and two pairs of their corresponding sides are parallel: $J(x, y) \parallel J(x', y')$ and $J(x, z) \parallel J(x', z')$. By the appropriate Desargues' theorem, we get $J(y, z) \parallel J(y', z')$.

As before, we define a map $\psi: A \to A$ as follows. Let $w \in A$, $w \neq p$. Suppose $w \notin J(p, x)$ (otherwise use $J(p, y)$). We define $\psi(w) = J(p, w) \cap L''$, where L'' is the line through x' parallel to $J(x, w)$. We also define $\psi(p) = p$. We assert that ψ is well defined and that the following are true (see Exercise 2):

(1) ψ is 1-1 and onto;
(2) if $\psi(w) = w$ for some $w \neq p$, then ψ is the identity map;
(3) ψ maps each line through p onto itself;
(4) ψ maps each line onto a parallel line.

The map ψ is called a *dilatation*. Again using the strength of our algebraic setting, we see that dilatations are easily treated (see Exercise 3).

Construction 3

Let L be a line in A. As indicated in Figure 2.9, we can define a map which sends x to x', y to y' such that $J(x, y) \cap J(x', y') \in L$, provided $J(x, y) \cap L \neq \varnothing$. Otherwise we take y' such that $J(x, y) \parallel J(x', y') \parallel L$. In either event, we take y' so that $J(y, y') \parallel J(x, x')$. This map is called a *perspectivity* of A with *axis* L. Properties of this map are dealt with in Exercise 4.

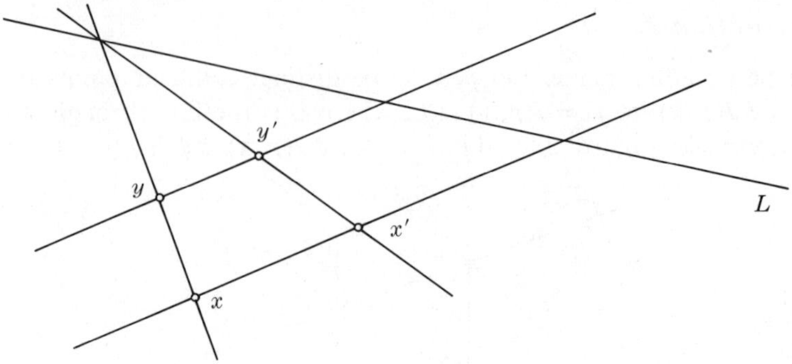

Figure 2.9

Summing up these three construction techniques, we see that they have some things in common. Each map is 1-1 and onto, and sends lines to lines. Such a map is called a *collineation.* The first two maps send each line to a parallel line; the third does not in general. The maps differ considerably in regard to their *fixed points*: the set of fixed points can be empty, a single point, or a line.

EXERCISES

1. Let ${}_DV$ be any vector space. For $v' \in V$ define a *translation* $t_{v'}: V \to V$ by $(v)t_{v'} = v + v'$, $\forall v \in V$.

   (a) Show that $t_{v'}$ is 1-1 and onto, for any v'.
   (b) Show that $(A)t_{v'}$ is an affine subspace if A is.
   (c) Show that $L((A)t_{v'}) = L(A)$ for any affine subspace A of V; hence $t_{v'}$ maps any affine subspace of V onto a parallel affine subspace of the same dimension.
   (d) Using the notation of Construction 1, show that $\varphi = t_{x'-x}$.

2. Verify the assertions of Construction 2 in the spirit of Construction 1 (that is, do not use algebraic techniques).

3. Let D be commutative. Let $p \in {}_DV$, and $\alpha \in D$. Define $f_\alpha: V \to V$ by $(x)f_\alpha = (1 - \alpha)p + \alpha x$, $\forall x \in V$.

   (a) Show that f_α is 1-1 and onto unless $\alpha = 0$.
   (b) Show that $(A)f_\alpha$ is an affine subspace if A is.
   (c) Show that $L((A)f_\alpha) = L(A)$ for any affine subspace A of V; hence f_α maps any affine subspace onto a parallel affine subspace of the same dimension.

(d) Using the notation of Construction 2, show that $\psi = f_\alpha$, where $x' = (1 - \alpha)p + \alpha x$.
(e) Discuss the special case where $V = \mathbb{R}^n$ and $p = O$.
(f) Note that there is a 1-1 correspondence $\alpha \leftrightarrow f_\alpha$ between elements of D and the set of dilatations; can more be said?
(g) Why does D need to be commutative?

4. Complete the details of Construction 3. In particular:

(a) show that the map is well defined (use a form of Desargues' theorem);
(b) show that the map is 1-1 and onto;
(c) show that L is exactly the set of fixed points.

SECTION 2.6
AFFINE MAPS

Motivated by the constructions in the preceding section, we set out to determine an algebraic characterization of the 1-1 maps of a vector space ${}_DV$ onto itself that carry lines to lines. Maps of this type will be called *collineations*. Observe that if dim $V = 1$, then any permutation of the vectors in V is such a map. We thus concern ourselves only with the cases dim $V \geq 2$. Since O plays a special role in ${}_DV$, we begin by considering maps that send O to O. As we shall see, it is easy to generalize by composing with translations.

Theorem 1

Let dim $V \geq 2$. Suppose $f: {}_DV \to {}_DV$ is a collineation such that $(O)f = O$. Then there exists an automorphism $\sigma: D \to D$ such that

$$(x + y)f = (x)f + (y)f,$$

and

$$(\alpha x)f = \alpha^\sigma (x)f, \; \forall x, y \in V, \forall \alpha \in D.$$

Proof

Let x, y be linearly independent vectors in ${}_DV$. The lines $J(O, y)$ and $J(x, x + y)$ are parallel. Their respective images $J(0, (y)f)$ and $J((x)f, (x + y)f)$ cannot intersect since f is 1-1. It will follow that these image lines are parallel provided we show that they lie in the same plane. But (see Figure 2.10), letting z be the midpoint of y and x, we know that $x + y \in J(O, z)$. Hence $(x + y)f \in J(O, (z)f) \subseteq J(O, (x)f, (y)f)$, and the lines $J(O, (y)f)$ and $J((x)f, (x + y)f)$

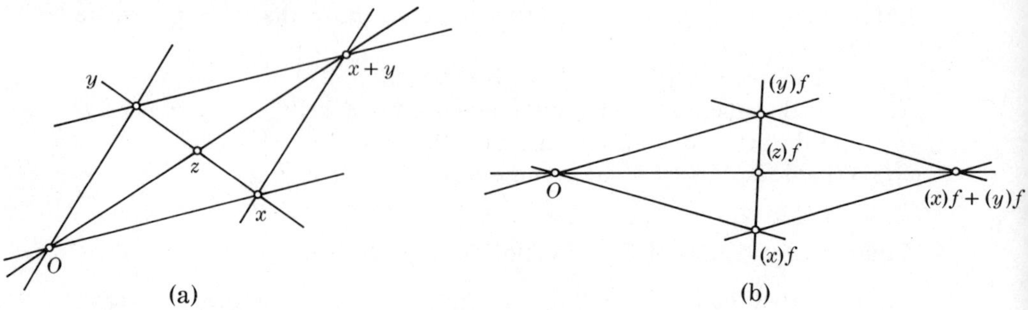

Figure 2.10

are parallel. Similarly $J(O, (x)f) \parallel J((y)f, (x+y)f)$. Therefore, $(O, (x)f, (x+y)f, (y)f)$ is a parallelogram. Also it is clear that $(O, (x)f, (x)f + (y)f, (y)f)$ is a parallelogram; hence $(x+y)f = (x)f + (y)f$. We need to establish that $(x+y)f = (x)f + (y)f$ in case x and y are linearly dependent. Of course, if $x = O$ or $y = O$ this is clear. Otherwise take $z \notin \langle x \rangle$. Then each of the pairs $\{(x+y), z\}$, $\{x, (y+z)\}$, and $\{y, z\}$ is linearly independent; hence

$$\begin{aligned}(x+y)f + (z)f = ((x+y)+z)f &= (x+(y+z))f \\ &= (x)f + (y+z)f = (x)f + (y)f + (z)f,\end{aligned}$$

giving

$$(x+y)f = (x)f + (y)f, \ \forall x, y \in V.$$

Once again, suppose x and y are linearly independent. Since, for $\alpha \in D$, we have $\alpha x \in J(O, x)$, it follows that $(\alpha x)f \in J(O, (x)f)$ or $(\alpha x)f = \alpha^\sigma \cdot (x)f$, for some function $\sigma: D \to D$, which appears to depend on x. Clearly $\alpha^\sigma = 0 \Leftrightarrow \alpha = 0$. Likewise $(\alpha y)f = \alpha^\tau (y)f$, for some function $\tau: D \to D$. However, letting $\alpha \neq 0$,

$$\begin{aligned}&J(\alpha x, \alpha y) \parallel J(x, y) \\ &\quad \Rightarrow J((\alpha x)f, (\alpha y)f) \parallel J((x)f, (y)f) \text{ (by the argument concerning Figure 2.10)} \\ &\quad \Rightarrow \langle \alpha^\sigma (x)f - \alpha^\tau (y)f \rangle = \langle (x)f - (y)f \rangle \\ &\quad \Rightarrow \alpha^\sigma (x)f - \alpha^\tau (y)f = \gamma((x)f - (y)f) \text{ (for some } \gamma \in D) \\ &\quad \Rightarrow \alpha^\sigma = \alpha^\tau.\end{aligned}$$

Thus the map $\sigma: D \to D$ is the same for x and y. Moreover, let $z \in V$. Then either $\{x, z\}$ or $\{y, z\}$ is linearly independent, and by repeating the above argument, we get

$$(\alpha z)f = \alpha^\sigma (z)f, \ \forall \alpha \in D.$$

We establish some properties of σ.

(1) $(\alpha + \beta)^\sigma = \alpha^\sigma + \beta^\sigma, \ \forall \alpha, \beta \in D.$

Proof

$$(\alpha^\sigma + \beta^\sigma)(x)f = \alpha^\sigma(x)f + \beta^\sigma(x)f = (\alpha x)f + (\beta x)f$$
$$= (\alpha x + \beta x)f = ((\alpha + \beta)x)f = (\alpha + \beta)^\sigma(x)f.$$

(2) $(\alpha\beta)^\sigma = \alpha^\sigma\beta^\sigma$.

Proof

$$\alpha^\sigma\beta^\sigma(x)f = \alpha^\sigma \cdot (\beta x)f = (\alpha\beta x)f = (\alpha\beta)^\sigma \cdot (x)f.$$

(3) σ maps D *onto* D and is 1-1.

Proof

Let $\beta \in D$. Since f is *onto*, and since f maps $J(O, x)$ onto $J(O, (x)f)$, $\beta(x)f = (\alpha x)f$ for some $\alpha \in D$. Thus $\alpha^\sigma = \beta$. Finally, σ is 1-1 since f is 1-1.

Properties 1, 2, and 3 establish that σ is an automorphism of D, and the proof of the theorem is complete.

DEFINITION 1 If σ is an automorphism of D, a map $f: V \to V$ such that $(x + y)f = (x)f + (y)f$, and $(\alpha x)f = \alpha^\sigma(x)f$, $\forall x, y \in V$, $\forall \alpha \in D$ is called a *semilinear map*. Note that a semilinear map need not be 1-1, although those arising from Theorem 1 are.

Corollary

Let $\dim V \geq 2$. Then any collineation is the composite of a semilinear map and a translation.

Proof

Let g be such a map. Put $f = g \cdot t_{-v}$, where $v = (O)g$. Then since translations are 1-1, are onto, and send lines to lines, f satisfies the hypotheses of Theorem 1, so that f is semilinear. But $g = f \cdot t_v$.

Actually, these maps we have been considering completely ignore an important aspect of the structure of ${}_DV$, namely, the notions of ratio and barycentric coordinates. For this reason we shall confine our attention to a more restricted class of maps—those that preserve ratio.

Proposition 1

Let f be semilinear and $v \in V$. If the map $g = f \cdot t_v$ preserves ratio (that is, $\alpha(x, y\colon z) = \alpha((x)g, (y)g\colon (z)g)$, $\forall x, y, z \in V$), then f is linear (that is, $\sigma =$ the identity automorphism).

Proof

Since translations preserve ratio, we may assume that f does also. Let $(x - z) = \alpha(y - z)$. Then $(x)f - (z)f = \alpha^\sigma((y)f - (z)f) \Rightarrow \alpha^\sigma = \alpha$. Since α can be arbitrary, we are done.

EXAMPLES

(1) If D is not commutative, then the dilatations f_α are in general composites of semilinear maps and translations. Specifically, letting O be the fixed point, define

$$(x)f_\alpha = \alpha x, \ \forall x \in V.$$

Note that $(\beta x)f_\alpha = \alpha\beta x \neq \beta(\alpha x)$ unless $\alpha\beta = \beta\alpha$. But $(\beta x)f_\alpha = \beta^\sigma \cdot (x)f_\alpha$, where $\beta^\sigma = \alpha\beta\alpha^{-1}$, and this map σ is an automorphism of D.

(2) Of course the field $\mathbb{R}$ has no nontrivial automorphisms, so linear and semilinear maps coincide for ${}_{\mathbb{R}}V$.

With this much background we can give an algebraic definition of an affine map that takes into account the affine structure of a vector space in precisely the way that linear maps acknowledged the linear structure of a space. We include in our definition those maps that are not 1-1 or onto.

DEFINITION 2 A function $f: {}_D V \rightarrow {}_D W$ is called an *affine map* if

$$(\alpha x + (1 - \alpha)y)f = \alpha(x)f + (1 - \alpha)(y)f, \ \forall x, y \in V, \quad \forall \alpha \in D.$$

Proposition 2

(a) If $f: V \rightarrow V$ is an affine map and f is 1-1 and onto, then f sends lines to lines.

(b) If $f: V \rightarrow V$ is 1-1, is onto, and preserves ratio, then f is an affine map.

Proof

(a) Let $z \in J(x, y)$. Then $z = \alpha x + (1 - \alpha)y$, and

$$(z)f = \alpha(x)f + (1 - \alpha)(y)f \in J((x)f, (y)f).$$

(b) Since f preserves ratio, f sends lines to lines; hence by Theorem 1 and Proposition 1, $f = g \cdot t_v$ with g linear. Hence

$$\begin{aligned}(\alpha x + (1 - \alpha)y)f &= \alpha(x)g + (1 - \alpha)(y)g + v \\ &= \alpha(x)g + (1 - \alpha)(y)g + \alpha v + (1 - \alpha)v \\ &= \alpha((x)g + v) + (1 - \alpha)((y)g + v) \\ &= \alpha(x)f + (1 - \alpha)(y)f.\end{aligned}$$

Although Proposition 2 makes the connection between 1-1 affine maps and their geometric motivation, the following theorem is far more significant to our development. We first need:

Lemma 1

(**a**) linear maps are affine maps;
(**b**) translations are affine maps;
(**c**) composites of affine maps are affine maps.

Proof

The proof is left as an exercise for the reader.

Theorem 2

Let $f{:}_D V \to {}_D W$. Then f is an affine map if and only if $f = g \cdot t_w$, for some linear map $g{:}V \to W$ and some $w \in W$.

Proof

$\boxed{\Rightarrow}$: Put $w = (O)f$, and define g by $g = f \cdot t_{-w}$. Then g is affine, by Lemma 1, and $(O)g = O$. For $\alpha \in D$, $x \in V$,

$$(\alpha x)g = (\alpha x + (1 - \alpha)O)g = \alpha(x)g + (1 - \alpha)(O)g = \alpha(x)g.$$

Now for $x, y \in V$, since $2^{-1} + 2^{-1} = 1$, we see that

$$\begin{aligned}(x + y)g &= (2(2^{-1}x + 2^{-1}y))g = 2(2^{-1}x + 2^{-1}y)g \\ &= 2[2^{-1}(x)g + 2^{-1}(y)g] = (x)g + (y)g.\end{aligned}$$

Thus g is linear, and $f = g \cdot t_w$.

$\boxed{\Leftarrow}$: This is a corollary to Lemma 1.

EXERCISES

1. Let $D = \mathbb{Z}_2$ and $V = {}_D D^n$, for any $n \geq 1$. Show that any *function* $f{:}V \to V$ is an affine map.

2. Let $D = \mathbb{Z}_2$.

   (a) Find an affine map $f{:}D^2 \to D^2$ that is not the composite of a linear map and a translation.
   (b) Find a 1-1 affine map $g{:}D^3 \to D^3$ that is not the composite of a linear map and a translation.

3. Let $A(V)$ be the set of all 1-1 affine maps of V onto V. Show that $A(V)$ is a group under composition.

4. Show that the factorization of an affine map into the composite of a linear map and a translation is unique. In other words, if $g \cdot t_{w'} = h \cdot t_{w''}$, with g and h linear, then $g = h$, and $t_{w'} = t_{w''}$.

5. Let f be an affine map from V to W, and A an affine subspace of V.
   (a) Show $(A)f$ is an affine subspace of W.
   (b) What is $L((A)f)$, the linear subspace of W associated with $(A)f$?

6. Show that parallelism is preserved by affine maps.

7. Let $f{:}_D V \to {}_D V$ be affine. If $\alpha_0, \cdots, \alpha_n \in D$, $\sum_{i=0}^{n} \alpha_i = 1$, then

$$(\alpha_0 a_0 + \alpha_1 a_1 + \cdots + \alpha_n a_n)f = \alpha_0(a_0)f + \cdots + \alpha_n(a_n)f, \; \forall a_0, a_1, \cdots, a_n \in V.$$

8. Let $T(V)$ be the set of all translations of $V \to V$. Show that $T(V)$ is an Abelian normal subgroup of $A(V)$.

DEFINITION Let $\mathscr{L}_1$, $\mathscr{L}_2$ be lattices. A function $\varphi{:}\mathscr{L}_1 \to \mathscr{L}_2$ is called a *lattice homomorphism* if $\varphi(x_1 \cap x_2) = \varphi(x_1) \cap \varphi(x_2)$ and $\varphi(x_1 \cup x_2) = \varphi(x_1) \cup \varphi(x_2)$, $\forall x_1, x_2 \in \mathscr{L}_1$. We call φ a *lattice monomorphism, epimorphism, isomorphism* if φ is 1-1, onto, both 1-1 and onto, respectively.

*9. (a) Show that each monomorphism $f{:}_D V \to {}_D W$ induces a lattice monomorphism $\hat{f}$ of $\mathscr{L}(V)$ into $\mathscr{L}(W)$ that is a lattice epimorphism or isomorphism respectively as f is a vector space epimorphism or isomorphism.
   (b) Do part (a) above for affine maps and $\mathscr{A}(V)$, $\mathscr{A}(W)$. Does f have to be monic in this case? [4.5]

10. Let Coll (V) be the set of all collineations of V onto V.
   (a) Show that Coll (V) is a group.
   (b) Show that $\{f \in \text{Coll}(V){:}(0)f = 0\} = H$ is a subgroup of Coll (V).
   (c) Show that $GL(V)$ is a normal subgroup of the group H in (b).

*11. Theorem 1 can be extended in the following manner. Let ${}_DV$, ${}_{D'}V'$ be vector spaces and define a *collineation* $f:V \to V'$ to be a 1-1 onto map sending lines of V onto lines on V'. Prove that if f is a collineation there is a division ring isomorphism $\sigma:D \to D'$ such that

(a) $(x + y)f = (x)f + (y)f$, $\forall x, y \in V$,
(b) $(\alpha x)f = \alpha^\sigma(x)f$, $\forall \alpha \in D$, $x \in V$.

We extend our previous terminology by referring to such a map f as *semi-linear*. [4.7]

SECTION 2.7

THE RELATIONSHIP BETWEEN THE AFFINE GROUP AND THE GENERAL LINEAR GROUP

Let V be a vector space. We wish to compare some of the groups of maps from V onto itself, such as $GL(V)$, $A(V)$, and $T(V)$. Some elementary observations can be made concerning these groups:

(1) $GL(V)$ and $T(V)$ are subgroups of $A(V)$;
(2) $T(V)$ is Abelian; $GL(V)$ is not;
(3) $GL(V) \cap T(V) = I$, the identity map on V;
(4) each element of $A(V)$ is a product (in a unique way) of an element of $GL(V)$ with an element of $T(V)$.

As a matter of fact, properties (3) and (4) bear some resemblance to the properties

(3′) $U \cap W = O$;
(4′) $v = u + w$ (uniquely) for some $u \in U$, and $w \in W$;

which are enjoyed by complements U and W in a vector space V. Our first theorem is an analogue of the theorem for vector spaces that asserts that the map $(u + w) \to u$ is an epimorphism of V onto U with kernel W where $V = U \oplus W$; hence $V/W \approx U$. The reader is invited to recall the appropriate definitions and facts from Chapter 1, Section 1.11.

Theorem 1

Let ${}_DV$ be a vector space. The function from $A(V)$ to $GL(V)$ given by $f \to \bar{f}$, where $f = \bar{f} \cdot t_v$, is a group epimorphism with kernel $T(V)$. Hence $A(V)/T(V) \approx GL(V)$.

Proof

Let $f = \bar{f} \cdot t_v$, $g = \bar{g} \cdot t_{v'}$ be in $A(V)$. Then

$$(x)f \cdot g = ((x)\bar{f} + v)g = (x)\bar{f}\bar{g} + (v)\bar{g} + v'.$$

Consequently $f \cdot g = (\bar{f} \cdot \bar{g}) \cdot t_{v''}$, where $v'' = (v)\bar{g} + v'$, showing that $\overline{(f \cdot g)} = \bar{f} \cdot \bar{g}$ so that $f \to \bar{f}$ is a group homomorphism. The function is onto since each element in $GL(V)$ is the image of *itself* under $f \to \bar{f}$. Finally, let f be in the kernel of the homomorphism, that is, $\bar{f} = I$, the identity on V. Then $f = \bar{f} \cdot t_v = I \cdot t_v = t_v \in T(V)$. On the other hand, if $f = t_v \in T(V)$, then $f = I \cdot t_v \Rightarrow \bar{f} = I$, and f is in the kernel. Thus $T(V)$ is the kernel of the map. The last assertion of the theorem is, of course, a consequence of the Fundamental Theorem of Homomorphisms for Groups.

The second result relating affine groups to linear groups involves a more intricate construction. It shows how to view an affine map on V as a linear map on a space V' whose dimension exceeds that of V by 1. One purpose for such a construction is that linear maps can be represented as matrices, and matrix theory can yield information about the maps. In fact we shall prove the theorem in what seems to be a very special case, $V = D^n$ and $V' = D^{n+1}$. However, the exercises that follow show that if $V \approx W$, then $A(V) \approx A(W)$, $GL(V) \approx GL(W)$, and $T(V) \approx T(W)$, so our results remain valid in a general setting.

Theorem 2

There is a group monomorphism taking $A(D^n)$ into $GL(D^{n+1})$.

Proof

We construct a function φ taking $A(D^n)$ into $GL(n+1, D)$, the group of invertible $(n+1) \times (n+1)$ matrices over D, which we know is isomorphic to $GL(D^{n+1})$. Suppose $B = \{(1, 0, \cdots, 0), \cdots, (0, \cdots, 1)\}$ is the standard basis of D^n. Let $f \in A(D^n)$. Then $f = \bar{f} \cdot t_v$, where $m(\bar{f}, B) = (\alpha_{ij}) \in D_n$ and $m(v, B) = (\gamma_1 \cdots \gamma_n)$. We define

$$\varphi(f) = \begin{pmatrix} \alpha_{11} & \cdots & \alpha_{1n} & 0 \\ \vdots & \cdots & \vdots & \vdots \\ \alpha_{n1} & \cdots & \alpha_{nn} & 0 \\ \gamma_1 & \cdots & \gamma_n & 1 \end{pmatrix} \in D_{n+1}.$$

Or, compactly,

$$\varphi(f) = \left(\begin{array}{c|c} m(\bar{f}, B) & 0 \\ \hline m(v, B) & 1 \end{array}\right).$$

Two facts are immediate: $\varphi(f)$ is invertible since $m(\bar{f}, B)$ is; and $\varphi(f) = I$ if and only if $f = I$, where I is the appropriate identity element.

To complete the proof it suffices to show that $\varphi(f \cdot g) = \varphi(f) \cdot \varphi(g)$. Suppose $g \in A(V)$, with $g = \bar{g} \cdot t_{v'}$, where $m(v', B) = (\gamma_1' \cdots \gamma_n')$. Then as we computed in the proof of Theorem 1, $f \cdot g = (\bar{f} \cdot \bar{g}) \cdot t_{v''}$, where $v'' = (v)\bar{g} + v'$. But then $m(v'', B) = m(v, B)m(\bar{g}, B) + m(v', B)$. Thus

$$\varphi(f \cdot g) = \left(\begin{array}{c|c} m(\bar{f}, B)m(\bar{g}, B) & 0 \\ \hline m(v, B)m(\bar{g}, B) + m(v', B) & 1 \end{array}\right),$$

which, by Exercise 4(b) of Chapter 1, Section 1.8, is precisely $\varphi(f) \cdot \varphi(g)$.

EXAMPLE In $\mathbb{R}^2$, take $B = \{(1, 0), (0, 1)\}$. Let $f = \bar{f} \cdot t_v$, where

$$m(\bar{f}, B) = \begin{pmatrix} \frac{\sqrt{2}}{2} & \frac{\sqrt{2}}{2} \\ -\frac{\sqrt{2}}{2} & \frac{\sqrt{2}}{2} \end{pmatrix},$$

and $v = (2, 3)$. Then f can be described as a rotation about the origin through 45 degrees followed by a translation which sends (0, 0) to (2, 3). The map given by

$$\varphi(f) = \begin{pmatrix} \frac{\sqrt{2}}{2} & \frac{\sqrt{2}}{2} & 0 \\ -\frac{\sqrt{2}}{2} & \frac{\sqrt{2}}{2} & 0 \\ 2 & 3 & 1 \end{pmatrix}$$

acts on $\mathbb{R}^3$ as follows. It maps each plane $z = \alpha$ (constant) onto itself by first rotating the plane about $(0, 0, \alpha)$ through 45 degrees and then translating by the vector $(2\alpha, 3\alpha, \alpha)$. In particular, the xy-plane is acted on by the rotation only, and the plane $z = 1$ is acted on in exactly the way $\mathbb{R}^2$ is acted on by f. The total effect on $\mathbb{R}^3$ is a rotation about the z-axis followed by a "shear transformation" parallel to the xy-plane moving each plane $z = k$ a distance proportional to its distance, k, from the xy-plane.

EXERCISES

1. Let ${}_DV \approx {}_DW$. Show that $A(V) \approx A(W)$, $GL(V) \approx GL(W)$, $T(V) \approx T(W)$. [*Hint*: Suppose $g: V \to W$ is an isomorphism. For each $f \in A(V)$, show that $g^{-1}fg \in A(W)$.]

2. (a) In the notation of Theorem 2, find the image of $T(D^n)$ in D_{n+1}.
   (b) We know that $T(V)$ is a normal subgroup of $A(V)$. (An easy proof is that $T(V)$ is the kernel of the map in Theorem 1.) Is the image of $T(D^n)$ normal in D_{n+1}? Explain your answer in the light of the Fundamental Theorem of Homomorphisms for Groups.

3. Find the set of all elements $f \in A(V)$ such that $f \cdot g = g \cdot f$, for all $g \in A(V)$. [*Hint*: Use Theorem 2.]

4. How many elements are there in $A(D^2)$ where $D = \mathbb{Z}_3$?

Chapter 3

Multilinear Algebra

The title of this chapter may sound foreboding to the reader, but the objectives are fairly simple. We are going to take a look at how the notions of length, area, volume, and angle occur in Euclidean geometry, and define certain functions from vector spaces to their scalars that, when restricted to the Euclidean plane or space, give length, area, and so on.

Up to now we have paid little or no price in additional difficulty for considering vector spaces over division rings and, as you will see in Chapter 4, we will have a real use for the added generality. Most of what we will do in this chapter could be done over a division ring. However, doing so would lead to many difficult technical problems, and since we have no applications in mind that would necessitate such generality, *we shall restrict our attention to those vector spaces* ${}_DV$ *where* D *is a field.*

SECTION 3.1
SEMIBILINEAR FORMS

Let us begin by considering the subject of distance and angles in $\mathbb{R}^2$. It is quite possible that in a course in analytic geometry you represented these by means of the inner product. If $x, y \in \mathbb{R}^2$, $x = (\alpha_1, \alpha_2)$, $y = (\beta_1, \beta_2)$, their inner product is given by $x \cdot y = \alpha_1\beta_1 + \alpha_2\beta_2$. Easy calculations show that the inner product is a map $\mathbb{R}^2 \times \mathbb{R}^2 \to \mathbb{R}$ satisfying:

(1) $(x + x') \cdot y = x \cdot y + x' \cdot y$;
(2) $x \cdot (y + y') = x \cdot y + x \cdot y'$;
(3) $(\alpha x) \cdot y = \alpha(x \cdot y)$;
(4) $x \cdot (\alpha y) = \alpha(x \cdot y)$;
(5) $x \cdot y = y \cdot x$;
(6) $x \cdot x \geq 0, \forall x \in \mathbb{R}^2$;
(7) $x \cdot x = 0 \Leftrightarrow x = O$.

You probably observed, as you read through this list of properties, that they are not logically independent. In the presence of (5), (1) and (2) are clearly equivalent, as are (3) and (4).

The length of $x \in \mathbb{R}^2$, denoted $\|x\|$, is then given by $\|x\| = \sqrt{x \cdot x}$ and the distance between x and y in $\mathbb{R}^2$ is given by $\|x - y\|$. The angle from x to y is the number θ such that $\cos\theta = (x \cdot y)/(\|x\|\ \|y\|)$. Since your knowledge of analytic geometry tells you how useful the inner product and its associated concepts are in $\mathbb{R}^2$, it is reasonable to ask whether an "inner" product exists on other vector spaces. The answer is both yes and no. If ${}_DV$ is a vector space and B is a basis, the map, $V \times V \to D$ defined by $x \cdot y = m(x, B)(m(y, B))^T$ for all $x, y \in V$, satisfies conditions (1) through (5) of the inner product, (1) through (3) being from properties of matrix multiplication and (4) and (5) following from our assumption that D is a field. Properties (6) and (7) are lost over arbitrary fields, however. Property (6) makes no sense over fields such as $\mathbb{Z}_p$ or $\mathbb{C}$ which are not ordered, and if we let $V = \mathbb{C}^2$, $B = \{(1, 0), (0, 1)\}$, $x = 1(1, 0) + i(0, 1)$, then we have $x \cdot x = (1 \quad i) \cdot \begin{pmatrix} 1 \\ i \end{pmatrix} = 1 + i^2 = 1 - 1 = 0$, and so (7) is not satisfied.

Properties (1)–(4) remain valid under a further generalization. If $\dim(V) = n$ and $(\alpha_{ij}) \in D_n$ we can define $V \times V \to D$ by

$$x \cdot y = m(x, B)(\alpha_{ij})(m(y, B))^T.$$

This function almost gives us the general situation we want to study in this section, but fails to cover one important example. If we let $V = \mathbb{C}^n$ and for $\alpha \in \mathbb{C}$ we denote by $\bar{\alpha}$ the complex conjugate of α, the map $V \times V \to \mathbb{C}$ given by $(\alpha_1, \cdots, \alpha_n) \cdot (\beta_1, \cdots, \beta_n) = \sum_{i=1}^n \alpha_i \bar{\beta}_i$ is known as a "Hermitian form" and is of some importance. Recalling that conjugation is an automorphism of $\mathbb{C}$, we see that all of the above functions are covered by the following definition.

DEFINITION 1 Let D be a field, $\sigma:D \to D$ an automorphism of D, and ${}_DV$ a vector space. *A σ-bilinear form* is a map $\Gamma:V \times V \to D$ such that:

(1) $\Gamma(x + x', y) = \Gamma(x, y) + \Gamma(x', y)$ for $x, x', y \in V$,
(2) $\Gamma(x, y + y') = \Gamma(x, y) + \Gamma(x, y')$ for $x, y, y' \in V$,
(3) $\Gamma(\alpha x, y) = \alpha \cdot \Gamma(x, y)$ for $\alpha \in D$, $x, y \in V$,
(4) $\Gamma(x, \alpha y) = \alpha^\sigma \cdot \Gamma(x, y)$ for $\alpha \in D$, $x, y \in V$.

In case σ is the identity map on D, Γ is called a *bilinear form*. The axioms for a σ-bilinear form say that for a fixed $y \in V$ the map $V \to D$ given by $x \to \Gamma(x, y)$ is a homomorphism, and for a fixed $x \in V$ the map $V \to D$ given by $y \to \Gamma(x, y)$ is a semilinear map. (See Chapter 2, Section 2.6.)

One of the basic facts about vector-space homomorphisms is that they are uniquely determined by what they do to a basis. This is also true of σ-bilinear forms; more precisely, we have the following:

Proposition 1

Let D be a field; $\sigma:D \to D$, an automorphism; ${}_DV$, a vector space; $B = \{x_1, \cdots, x_n\}$ a basis of V; and $\Gamma, \Delta:V \times V \to D$, σ-bilinear forms. Then $\Gamma = \Delta \Leftrightarrow \Gamma(x_i, x_j) = \Delta(x_i, x_j)$ for all i, j, $1 \le i \le n$, $1 \le j \le n$.

Proof

$\boxed{\Rightarrow}$: This implication is a consequence of the definition of equality of maps.

$\boxed{\Leftarrow}$: Let $x, y \in V$. Then for suitable $\alpha_1, \cdots, \alpha_n$, $\beta_1, \cdots, \beta_n \in D$, $x = \alpha_1 x_1 + \cdots + \alpha_n x_n$, $y = \beta_1 x_1 + \cdots + \beta_n x_n$, and

$$\begin{aligned}\Gamma(x, y) &= \Gamma\left(\sum_{i=1}^n \alpha_i x_i, \sum_{j=1}^n \beta_j x_j\right)\\ &= \sum_{i=1}^n \alpha_i \Gamma\left(x_i, \sum_{j=1}^n \beta_j x_j\right)\\ &= \sum_{i=1}^n \sum_{j=1}^n \alpha_i \beta_j^\sigma \Gamma(x_i, x_j)\\ &= \sum_{i=1}^n \sum_{j=1}^n \alpha_i \beta_j^\sigma \Delta(x_i, x_j)\\ &= \Delta\left(\sum_{i=1}^n \alpha_i x_i, \sum_{j=1}^n \beta_j x_j\right)\\ &= \Delta(x, y).\end{aligned}$$

Proposition 1 should be viewed as the σ-bilinear form analogue of Proposition 1, in Chapter 1, Section 1.7, and Theorem 1 should be viewed as the σ-bilinear form analogue of Theorem 1 of Chapter 1, Section 1.7 and Proposition 5 of Chapter 1, Section 1.8. Before stating Theorem 1 we need a notational device. If $\sigma: D \to D$ is an automorphism and $(\alpha_{ij}) \in D_{n\times m}$ we shall define

$$(\alpha_{ij})^\sigma = (\alpha_{ij}{}^\sigma).$$

Theorem 1

Let D be a field, $\sigma: D \to D$ an automorphism, ${}_DV$ a vector space, and $B = \{x_1, \cdots, x_n\}$ a basis of V. Then

(1) If $\Gamma: V \times V \to D$ is a σ-bilinear form there exists a unique matrix, denoted $m(\Gamma, B)$, in D_n such that

$$\Gamma(x, y) = m(x, B)m(\Gamma, B)(m(y, B)^T)^\sigma, \text{ for all } x, y \in V.$$

(2) If $(\alpha_{ij}) \in D_n$ there exists a unique σ-bilinear form $\Gamma: V \times V \to D$ such that $\Gamma(x, y) = m(x, B)(\alpha_{ij})(m(y, B)^T)^\sigma$

Proof

(1) Let $m(\Gamma, B) = (\alpha_{ij})$, where $\alpha_{ij} = \Gamma(x_i, x_j)$. Then if $x = \beta_1 x_1 + \cdots + \beta_n x_n$ and $y = \delta_1 x_1 + \cdots + \delta_n x_n$, we have

$$\begin{aligned}
m(x, B)m(\Gamma, B)(m(y, B)^T)^\sigma &= (\beta_1 \cdots \beta_n)\begin{pmatrix} \alpha_{11} & \cdots & \alpha_{1n} \\ \vdots & & \vdots \\ \alpha_{n1} & \cdots & \alpha_{nn} \end{pmatrix}\begin{pmatrix} \delta_1{}^\sigma \\ \vdots \\ \delta_n{}^\sigma \end{pmatrix} \\
&= \left(\sum_{k=1}^n \beta_k \alpha_{k1} \cdots \sum_{k=1}^n \beta_k \alpha_{kn}\right)\begin{pmatrix} \delta_1{}^\sigma \\ \vdots \\ \delta_n{}^\sigma \end{pmatrix} \\
&= \sum_{l=1}^n \left(\sum_{k=1}^n \beta_k \alpha_{kl} \delta_l{}^\sigma\right) \\
&= \sum_{l=1}^n \sum_{k=1}^n \beta_k \delta_l{}^\sigma \Gamma(x_k, x_l) \\
&= \Gamma\left(\sum_{k=1}^n \beta_k x_k, \sum_{l=1}^n \delta_l x_l\right) \\
&= \Gamma(x, y).
\end{aligned}$$

(2) That the function $\Gamma(x, y)$ defined by

$$\Gamma(x, y) = m(x, B)(\alpha_{ij})(m(y, B)^T)^\sigma$$

is a σ-bilinear form is Exercise 2. Its uniqueness follows from Proposition 1 and the fact that

$$m(x_i, B)(\alpha_{ij})(m(x_j, B)^T)^\sigma = \alpha_{ij}.$$

So far we have established a relation between σ-bilinear forms and matrices. We will see that this is a fruitful relationship, for our knowledge of matrices will yield information about σ-bilinear forms, and theorems about σ-bilinear forms will also give information about matrices.

As a final point in the relation between σ-bilinear forms and matrices we want to discuss the relationship between matrices of a σ-bilinear form with respect to different bases. Let $\Gamma: V \times V \to D$ be a σ-bilinear form, $B = \{x_1, \cdots, x_n\}$, $C = \{y_1, \cdots, y_n\}$ two bases of V, $m(\Gamma, B) = (\alpha_{ij})$, and $m(\Gamma, C) = (\beta_{ij})$. Suppose $x_i = \sum_{k=1}^n \tau_{ik} y_k$. Then

$$\Gamma(x_i, x_j) = \Gamma\left(\sum_{k=1}^n \tau_{ik} y_k, \sum_{l=1}^n \tau_{jl} y_l\right)$$

$$= \sum_{k=1}^n \sum_{l=1}^n \tau_{ik} \tau_{jl}{}^\sigma \Gamma(y_k, y_l).$$

Let $1: V \to V$ be the identity map. Then

$$(\tau_{ij}) = m(1, B, C)$$

and

$$m(\Gamma, B) = m(1, B, C) m(\Gamma, C)(m(1, B, C)^T)^\sigma.$$

EXERCISES

1. In each of the following assume that the matrix is formed relative to the standard basis.
   (a) Let $V = \mathbb{C}^2$, and σ be complex conjugation. Show that the σ-bilinear form Γ whose matrix is $\begin{pmatrix} 1+i & i \\ 1 & 2+2i \end{pmatrix}$ has the property that $\Gamma(x, y) = i\Gamma(y, x)^\sigma$, $\forall x, y \in V$.
   (b) Let $V = \mathbb{Z}_3{}^2$. Show that the bilinear form Γ whose matrix is $\begin{pmatrix} 0 & 1 \\ 2 & 0 \end{pmatrix}$ has the property that $\Gamma(x, y) = -\Gamma(y, x)$, $\forall x, y \in V$.
   (c) If $V = \mathbb{Z}_3{}^2$, show that the bilinear form Γ whose matrix is $\begin{pmatrix} 0 & 2 \\ 2 & 0 \end{pmatrix}$ has the property that $\Gamma(x, y) = \Gamma(y, x)$, $\forall x, y \in V$.

*2. Let ${}_DV$ be a vector space, $\sigma: D \to D$ an automorphism, $(\alpha_{ij}) \in D_n$, $B = \{x_1, \cdots, x_n\}$ a basis of V. Show that the map $\Gamma: V \times V \to D$ defined by

$$\Gamma(x, y) = m(x, B)(\alpha_{ij})(m(y, B)^T)^\sigma$$

is a σ-bilinear form. [3.2]

3. Let $V = \mathbb{R}^3$, $B = \{e_1, e_2, e_3\}$ the standard basis of $\mathbb{R}^3$, $\Gamma: V \times V \to \mathbb{R}$ the bilinear form such that

$$m(\Gamma, B) = \begin{pmatrix} 1 & 0 & 1 \\ 0 & 1 & 0 \\ 1 & 1 & 1 \end{pmatrix}.$$

(a) If $x = (\alpha_1, \alpha_2, \alpha_3)$, $y = (\beta_1, \beta_2, \beta_3) \in \mathbb{R}^3$, compute $\Gamma(x, y)$. Does $\Gamma(x, y) = \Gamma(y, x)$?
(b) Find $\{x \in \mathbb{R}^3 : \Gamma(x, y) = 0 \text{ for all } y \in \mathbb{R}^3\}$, and show that it is a subspace.
(c) Find $\{y \in \mathbb{R}^3 : \Gamma(x, y) = 0 \text{ for all } x \in \mathbb{R}^3\}$, and show that it is a subspace.
(d) Let $y_0 = (1, 1, 1)$, and show that $\{x \in \mathbb{R}^3 : \Gamma(x, y_0) = 1\}$ is a 2-dimensional affine subspace of $\mathbb{R}^3$.
(e) Find the matrix of Γ with respect to the basis

$$C = \{(1, 1, 0), (0, 1, 1), (-1, 0, -1)\}.$$

4. Let $V = C([0, 1])$. Show that the map $\Gamma: V \times V \to \mathbb{R}$ defined by

$$\Gamma(f, g) = \int_0^1 f(t)g(t)\, dt$$

is a bilinear form.

SECTION 3.2
SINGULARITY AND PERPENDICULARITY

The bilinear form of Exercise 2 of Section 3.1 demonstrates two pathologies that can occur in bilinear and σ-bilinear forms, but that do not occur in the inner product on $\mathbb{R}^2$. One is singularity, that is, there can be $x \neq O$ such that $\Gamma(x, y) = 0$ for all y; and y such that $\Gamma(x, y) = 0$ for all x. The other pathology is that perpendicularity may not be well defined. To be precise, in $\mathbb{R}^2$ the vectors x and y are perpendicular if $x \cdot y = 0$. Over a vector space ${}_DV$ with σ-bilinear form $\Gamma: V \times V \to D$, we might try to define x perpendicular to y

if $\Gamma(x, y) = 0$. However, x perpendicular to y should, if perpendicularity is to have any relation to our intuition, imply y perpendicular to x, that is, $\Gamma(x, y) = 0 \Rightarrow \Gamma(y, x) = 0$. But this does not always happen. We shall now study these phenomena in some detail.

DEFINITION 1 Let ${}_DV$ be a vector space and $\Gamma: V \times V \to D$ a σ-bilinear form. The left kernel of Γ, denoted LK (Γ), is defined by

$$\text{LK}(\Gamma) = \{x \in V : \Gamma(x, y) = 0 \text{ for all } y \in V\},$$

and the right kernel of Γ, denoted RK (Γ), is defined by

$$\text{RK}(\Gamma) = \{y \in V : \Gamma(x, y) = 0 \text{ for all } x \in V\}.$$

Theorem 1

If $\Gamma: V \times V \to D$ is a σ-bilinear form, then

(a) LK (Γ) and RK (Γ) are subspaces of V;
(b) dim (LK(Γ)) = dim (RK (Γ)).

Proof

(a) If $y_1, y_2 \in \text{RK}(\Gamma)$ and $x \in V$, then

$$\Gamma(x, y_1 + y_2) = \Gamma(x, y_1) + \Gamma(x, y_2) = 0 + 0 = 0$$

so $y_1 + y_2 \in \text{RK}(\Gamma)$. If $\alpha \in D$, then

$$\Gamma(x, \alpha y_1) = \alpha^\sigma \Gamma(x, y_1) = \alpha^\sigma \cdot 0 = 0$$

so $\alpha y_1 \in \text{RK}(\Gamma)$. LK $(\Gamma) < V$ is proved in like manner.

(b) Let $B = \{x_1, \cdots, x_n\}$ be a basis of V and $\alpha_{ij} = \Gamma(x_i, x_j)$. Hence $\Gamma(x, y) = m(x, B)(\alpha_{ij})(m(y, B)^T)^\sigma$ for all $x, y \in V$. Let $u_1, \cdots, u_n$ be the columns of (α_{ij}) considered as vectors in D^n. Now

$$\begin{aligned} y \in \text{RK}(\Gamma) &\Leftrightarrow \Gamma(x, y) = 0 \text{ for all } x \in V \\ &\Leftrightarrow \Gamma(x_i, y) = 0 \text{ for } i = 1, \cdots, n \\ &\Leftrightarrow (\alpha_{ij})(m(y, B)^T)^\sigma = 0 \end{aligned}$$

(that is, if $y = \beta_1 x_1 + \cdots + \beta_n x_n$)

$$\Leftrightarrow ({\beta_1}^\sigma, \cdots, {\beta_n}^\sigma) \in \mathscr{D}(\{u_1, \cdots, u_n\}).$$

(See Chapter 1, Section 1.3, Exercise 6.) It follows that dim RK (Γ) = $[n -$ column rank of $(\alpha_{ij})]$. A similar analysis of the left kernel shows dim (LK (Γ) = $[n -$ row rank of $(\alpha_{ij})]$, and by Chapter 1, Section 1.9, Proposition 4, dim LK (Γ) = dim RK (Γ). An alternate proof will be given in Exercise 3.

DEFINITION 2 If $\Gamma : V \times V \to D$ is a σ-bilinear form and LK $(\Gamma) = \text{RK}(\Gamma) = 0$ then Γ is called *nonsingular*.

We now identify those σ-bilinear forms that can be used to discuss perpendicularity, restricting our attention to the nonsingular case. The singular forms will be discussed in the exercises.

Theorem 2

Let ${}_DV$ be a vector space, $\Gamma : V \times V \to D$ a nonsingular σ-bilinear form, and $n = \dim(V) \geq 2$. Then $(\Gamma(x, y) = 0 \Rightarrow \Gamma(y, x) = 0$ for $x, y \in V) \Leftrightarrow$ either (1) $\Gamma(x, y) = \Gamma(y, x)$ for all $x, y \in V$ and $\sigma = 1$ or (2) $\Gamma(x, y) = -\Gamma(y, x)$ for all $x, y \in V$ and $\sigma = 1$ or (3) there exists $\alpha \in D$ such that $\Gamma(x, y) = \alpha(\Gamma(y, x))^\sigma$, $\sigma \neq 1$, $\sigma^2 = 1$.

Proof

In order to simplify notation we will write $x \cdot y$ for $\Gamma(x, y)$.

$\boxed{\Leftarrow}$: This direction is clear.

$\boxed{\Rightarrow}$: There are two cases to consider.

Case 1

$x \cdot x = 0$ for all $x \in V$. Then for $x, y \in V$,

$$\begin{aligned} 0 &= (x + y) \cdot (x + y) = x \cdot x + x \cdot y + y \cdot x + y \cdot y \\ &= x \cdot y + y \cdot x, \end{aligned}$$

and $x \cdot y = -y \cdot x$ for all $x, y \in V$. Given $y \in V$, $y \neq 0$, since Γ is nonsingular there exists $x' \in V$ such that $x' \cdot y = \alpha \neq 0$. Letting $x = \alpha^{-1}x'$ we have $x \cdot y = 1$ and so for any $\beta \in D$,

$$\begin{aligned} \beta^\sigma &= \beta^\sigma \cdot 1 = \beta^\sigma(x \cdot y) = x \cdot (\beta y) \\ &= -(\beta y) \cdot x = -\beta(y \cdot x) = \beta(x \cdot y) = \beta, \end{aligned}$$

and hence σ is the identity automorphism.

Case 2

There exists an $x_0 \in V$ such that $x_0 \cdot x_0 = \alpha_0 \neq 0$. The first step is to show that for any $x \in V$ there exists $\alpha_x \in D$ such that $x \cdot y = \alpha_x(y \cdot x)^\sigma$ for all $y \in V$. It should be noted that if α_x exists it must be unique, for if $x \cdot y = \alpha(y \cdot x)^\sigma$ for all $x, y \in V$, choose y such that $x \cdot y \neq 0$ (and hence by hypothesis $y \cdot x \neq 0$) and then $\alpha(y \cdot x)^\sigma = \alpha_x(y \cdot x)^\sigma \Rightarrow \alpha = \alpha_x$.

To establish the existence of α_x, let $y, z \in V$ be any vectors such that $y \cdot x \neq 0$ and $z \cdot x \neq 0$. Then

$$0 = 1 - 1 = \frac{y \cdot x}{y \cdot x} - \frac{z \cdot x}{z \cdot x} = \left(\frac{1}{y \cdot x} y - \frac{1}{z \cdot x} z\right) \cdot x$$

$$\Rightarrow 0 = x\left(\frac{1}{y \cdot x} y - \frac{1}{z \cdot x} z\right) = \frac{x \cdot y}{(y \cdot x)^\sigma} - \frac{x \cdot z}{(x \cdot z)^\sigma}$$

so

$$\frac{x \cdot y}{(y \cdot x)^\sigma} = \frac{x \cdot z}{(z \cdot x)^\sigma}.$$

If we set $\alpha_x = (x \cdot z)/(z \cdot x)^\sigma$ then $x \cdot y = \alpha_x(y \cdot x)$ for all $y \in V$ (our argument establishes the equality if $x \cdot y \neq 0$ and if $x \cdot y = 0$ it is obvious).

Next we claim that if $x, z \in V$ then $\alpha_x = \alpha_z$. First assume that x and z are linearly independent. Let $K_x = \{y \in V : y \cdot x = 0\}$, $K_z = \{y \in V : y \cdot z = 0\}$. K_x is the kernel of the map $V \to D$ given by $y \to y \cdot x$. Since $x \notin \mathrm{RK}(\Gamma)$, this map is not 0, so its image is D which has dimension 1; therefore $\dim(K_x) = n - 1$. Similarly $\dim(K_z) = n - 1$. $K_x \subseteq K_z \Rightarrow K_x = K_z$ so there exists $w \in V$ such that $V = K_x \oplus \langle w \rangle = K_z \oplus \langle w \rangle$. If $w \cdot x = \gamma_x$ and $w \cdot z = \gamma_z$ then $\gamma_x \neq 0$, $\gamma_z \neq 0$ and we have nonzero $\lambda_x, \lambda_z \in D$ such that $\lambda_x^\sigma = \gamma_x^{-1}$, $\lambda_z^\sigma = \gamma_z^{-1}$. If $y \in V$ then $y = y' + \tau w$ where $y' \in K_x = K_z$ and $\tau \in D$, and

$$\begin{aligned} y \cdot (\lambda_x x - \lambda_z z) &= (y' + \tau w) \cdot (\lambda_x x - \lambda_z z) \\ &= \tau \gamma_x^{-1} w \cdot x - \tau \gamma_z^{-1} w \cdot z = \tau - \tau = 0. \end{aligned}$$

Since y was arbitrary, this says $(\lambda_x x - \lambda_z z) \in \mathrm{RK}(\Gamma) = 0$, contradicting the linear independence of x and z. It follows that $K_x \nsubseteq K_z$, and we can find u such that $u \cdot x = 1$ and $u \cdot z = 0$. Similarly $K_z \nsubseteq K_x$ and there exists $v \in V$ such that $v \cdot x = 0$ and $v \cdot z = 1$. Hence

$$\begin{aligned} \alpha_x - \alpha_z &= \alpha_x \cdot 1 - \alpha_z \cdot 1 = \alpha_x(1)^\sigma - \alpha_z(1)^\sigma \\ &= \alpha_x(u \cdot x)^\sigma - \alpha_z(v \cdot z)^\sigma = x \cdot u - z \cdot v \\ &= x \cdot u - x \cdot v + z \cdot u - z \cdot v \text{ (note that } x \cdot v = z \cdot u = 0) \\ &= x \cdot (u - v) + z \cdot (u - v) = (x + z) \cdot (u - v) \\ &= \alpha_{x+z}((u - v) \cdot (x + z))^\sigma = \alpha_{x+z}(u \cdot x + u \cdot z - v \cdot x - v \cdot z)^\sigma \\ &= \alpha_{x+z}(1 - 1)^\sigma = 0. \end{aligned}$$

If x, z are linearly dependent, since $\dim V \geq 2$ there exists $w \in V$ such that $\{x, w\}$ and $\{z, w\}$ are linearly independent sets and $\alpha_x = \alpha_w = \alpha_z$.

We now have a unique $\alpha \in D$ such that $x \cdot y = \alpha(y \cdot x)^\sigma$ for all $x, y \in V$. Using the x_0 such that $x_0 \cdot x_0 = \alpha_0 \neq 0$, we have $x_0 \cdot x_0 = \alpha(x_0 \cdot x_0)^\sigma \Rightarrow \alpha_0 = \alpha\alpha_0^\sigma$ or $\alpha = \alpha_0/\alpha_0^\sigma$. If σ is the identity automorphism we have $\alpha = 1$ and $x \cdot y = y \cdot x$ for all $x, y \in V$, which is conclusion (1). If σ is not the identity, then

$$x \cdot y = \alpha(y \cdot x)^\sigma = \alpha(\alpha(x \cdot y)^\sigma)^\sigma$$
$$= \alpha\alpha^\sigma(x \cdot y)^{\sigma^2}.$$

Since we can find $x_1, y_1 \in V$ such that $x_1 \cdot y_1 = 1$, we have $\alpha\alpha^\sigma = 1$ and for any $\beta \in D$,

$$\beta = \beta 1 = \beta(x_1 \cdot y_1) = (\beta x_1 \cdot y_1) = ((\beta x_1) \cdot y_1)^{\sigma^2} = \beta^{\sigma^2}$$

Hence $\sigma^2 = 1$ and part (3) follows.

DEFINITION 3 Let ${}_DV$ be a vector space and $\Gamma: V \times V \to D$ a σ-bilinear form. If $\Gamma(x, y) = \Gamma(y, x)$ for all $x, y \in V$, Γ is said to be *symmetric*. If $\Gamma(x, y) = -\Gamma(y, x)$ for all $x, y \in V$, Γ is said to be *skew-symmetric*. If $\sigma^2 = 1$, $\sigma \neq 1$, and $\Gamma(x, y) = \alpha(\Gamma(y, x))^\sigma$ for all $x, y \in V$, Γ is said to be *Hermitian*. A vector space together with a symmetric, skew-symmetric, or Hermitian form is called respectively a symmetric, skew-symmetric, or Hermitian space.

EXERCISES

1. Prove Theorem 2 where Γ is allowed to be singular but $\dim(V/\mathrm{LK}(\Gamma)) \geq 2$. [*Hint*: Note that $\Gamma(x, y) = 0 \Rightarrow \Gamma(y, x) = 0$ says, among other things, that $\mathrm{LK}(\Gamma) = \mathrm{RK}(\Gamma)$, and let M be a complement of them, that is, $V = M \oplus \mathrm{LK}(\Gamma) = M \oplus \mathrm{RK}(\Gamma)$. Consider $\Delta: M \times M \to D$ defined by $\Delta(x, y) = \Gamma(x, y)$ for $x, y \in M$.]

2. Let ${}_DV$ be a vector space and $B = \{x_1, \cdots, x_n\}$ a basis.
   (a) If $\Gamma: V \times V \to D$ is a bilinear form, show that Γ is symmetric $\Leftrightarrow$ $m(\Gamma, B) = m(\Gamma, B)^T$.
   (b) If $\Gamma: V \times V \to D$ is a bilinear form, show that Γ is skew-symmetric $\Leftrightarrow m(\Gamma, B) = -m(\Gamma, B)^T$.
   (c) Let $\Gamma: V \times V \to D$ be a σ-bilinear form, $\sigma^2 = 1$. Show that Γ is Hermitian $\Leftrightarrow$ there is an $\alpha \in D$, $\alpha\alpha^\sigma = 1$ such that $m(\Gamma, B) = \alpha \cdot (m(\Gamma, B)^T)^\sigma$ where $\alpha \cdot (m(\Gamma, B)^T)^\sigma = (\sum_{i=1}^n \alpha e_{ii})(m(\Gamma, B)^T)^\sigma$.

3. This is to be an alternate proof of the fact that $\dim \mathrm{LK}(\Gamma) = \dim \mathrm{RK}(\Gamma)$. Let $V = M \oplus \mathrm{RK}(\Gamma)$, with $\{x_1, \cdots, x_k\}$ a basis of M and $\{x_1, \cdots, x_k, x_{k+1}, \cdots, x_n\}$ a basis of V. Show that there exist $y_1, \cdots, y_k \in V$ linearly independent such that $\Gamma(y_i, x_i) = 1$, $i = 1, \cdots, k$, $\Gamma(y_i, x_j) = 0$, $i \neq j$. Show that $\langle y_1, \cdots, y_k \rangle \cap \mathrm{LK}(\Gamma) = 0$ so $\dim \mathrm{LK}(\Gamma) \leq \dim \mathrm{RK}(\Gamma)$. An analogous construction yields the opposite inequality.

SECTION 3.3
THE STRUCTURE OF SYMMETRIC, SKEW-SYMMETRIC, AND HERMITIAN SPACES

DEFINITION 1 If $\Gamma: V \times V \to D$ is a symmetric, skew-symmetric, or Hermitian form, and $S, T \subseteq V$, we say that S *is perpendicular to* T, denoted $S \perp T$, if $\Gamma(s, t) = 0$ for all $s \in S, t \in T$. We denote by $S^{\perp}$, $\{x \in V: \Gamma(x, s) = 0$ for all $s \in S\}$. We say that *V is the orthogonal sum of the subspaces* $U_1, \cdots, U_n$, denoted $V = U_1 \perp \cdots \perp U_n$ provided $V = U_1 \oplus \cdots \oplus U_n$ and $U_i \perp \sum_{j=1, j \neq i}^{n} U_j$ for $i = 1, \cdots, n$.

The results that follow show how each of the three types of spaces (symmetric, skew-symmetric, Hermitian) can be decomposed into the orthogonal sum of subspaces of dimension 1 or 2. These theorems are generalizations of the fact that in $\mathbb{R}^2$ and $\mathbb{R}^3$ you can choose a basis of mutually perpendicular vectors of length 1.

Theorem 1

If ${}_DV$ is finite dimensional and $\Gamma: V \times V \to D$ is a symmetric form and $2^{-1} \in D$, then there exists a basis $B = \{x_1, \cdots, x_n\}$ of V such that $m(\Gamma, B)$ is diagonal, or alternately a finite dimensional symmetric space is an orthogonal sum of lines.

Proof

The proof is by induction on the dimension of V. If $\dim(V) = 1$ the result is obvious. Suppose the statement is true for every space of dimension $n - 1$ and V is of dimension n. If $\Gamma(x, x) = 0$ for all $x \in V$ then $\Gamma(x, y) = -\Gamma(y, x)$ (see proof of Theorem 2, Section 3.2). But $\Gamma(x, y) = \Gamma(y, x)$ by hypothesis, so $2\Gamma(y, x) = 0 \Rightarrow \Gamma(y, x) = 0$ for all $x, y \in V$ and any basis will do. If not, there is an $x_1 \in V$ such that $\Gamma(x_1, x_1) = \alpha_1 \neq 0$. Let $K_{x_1} = \{y \in V: \Gamma(y, x_1) = 0\}$, $\dim(K_{x_1}) = n - 1$ and $x_1 \notin K_{x_1}$ so $V = \langle x_1 \rangle \perp K_{x_1}$. Γ restricted to $K_{x_1} \times K_{x_1}$ is a symmetric bilinear form and so by induction hypothesis there exists a basis $\{x_2, \cdots, x_n\}$ of K_{x_1} such that $\Gamma(x_i, x_j) = 0$, $i \neq j$, $2 \leq i, j \leq n$. Since $\Gamma(x_1, x_i) = \Gamma(x_i, x_1) = 0$, $2 \leq i \leq n$, $\{x_1, \cdots, x_n\}$ is the required basis.

Since a diagonal matrix, that is, an (α_{ij}) such that $i \neq j \Rightarrow \alpha_{ij} = 0$, obviously has the property that $(\alpha_{ij})^T = (\alpha_{ij})$, this theorem is the same as the classical matrix theory result that asserts, for $A \in D_n$, $A = A^T \Leftrightarrow$ there exists a nonsingular $B \in D_n$ such that BAB^T is diagonal.

In order to handle skew-symmetric spaces we need a definition.

DEFINITION 2 Let $\Gamma:V \times V \to D$ be a bilinear form. A 2-dimensional subspace H of V is called a *hyperbolic plane* if it has a basis $\{x_1, x_2\}$ such that $\Gamma(x_1, x_1) = \Gamma(x_2, x_2) = 0$, $\Gamma(x_1, x_2) = -\Gamma(x_2, x_1) = 1$. The set $\{x_1, x_2\}$ is called a *hyperbolic pair*.

Theorem 2

Let ${}_DV$ be a finite dimensional vector space and $\Gamma:V \times V \to D$ a nonsingular skew-symmetric form, and suppose $2^{-1} \in D$. Then V is an orthogonal sum of hyperbolic planes and the dimension of V is an even integer.

Proof

Let $n = \dim(V)$. If $n = 1$ then $V = \langle x \rangle$, and $\Gamma(x, x) = -\Gamma(x, x) = 0$ so $\Gamma(\alpha x, \beta x) = \alpha\beta\Gamma(x, x) = 0$ for all $\alpha x, \beta x \in V$ contradicting the nonsingularity of Γ, so $\dim V \neq 1$. If $\dim V = 2$ let $x_1 \in V$, $x_1 \neq O$. Since Γ is nonsingular, there exists $x_2 \in V$ such that $\Gamma(x_1, x_2) = 1$. Then $\Gamma(x_2, x_1) = -\Gamma(x_1, x_2) = -1$ and $\Gamma(x_1, x_1) = \Gamma(x_2, x_2) = 0$. The set $\{x_1, x_2\}$ is linearly independent, for if not $x_2 = \alpha x_1$ for some $\alpha \in D \Rightarrow \Gamma(x_1, x_2) = 0$. Let $\dim V = n \geq 3$ and assume that for any vector space W of dimension less than n with a nonsingular skew-symmetric form, W is an orthogonal sum of hyperbolic planes and of even dimension. Choose $x_1 \in V$, $x_1 \neq 0$. As in the case of the plane there is an $x_2 \in V$ such that $\Gamma(x_1, x_2) = -\Gamma(x_2, x_1) = 1$, $\Gamma(x_1, x_1) = \Gamma(x_2, x_2) = 0$, and $\{x_1, x_2\}$ is linearly independent. As usual let

$$K_{x_1} = \{y \in V:\Gamma(y, x_1) = 0\},\ K_{x_2} = \{y \in V:\Gamma(y, x_2) = 0\};$$

$x_1 \in K_{x_1}$ but $x_1 \notin K_{x_2}$ so $V = K_{x_1} + K_{x_2}$. It follows that

$$\begin{aligned}\dim(K_{x_1} + K_{x_2}) + \dim(K_{x_1} \cap K_{x_2}) &= \dim(K_{x_1}) + \dim(K_{x_2})\\ &\Rightarrow n + \dim(K_{x_1} \cap K_{x_2}) = n - 1 + n - 1,\end{aligned}$$

and $\dim(K_{x_1} \cap K_{x_2}) = n - 2$. Since $\langle x_1, x_2 \rangle \cap (K_{x_1} \cap K_{x_2}) = O$, we have $V = \langle x_1, x_2 \rangle \oplus (K_{x_1} \cap K_{x_2})$. Γ restricted to $(K_{x_1} \cap K_{x_2}) \times (K_{x_1} \cap K_{x_2})$ is a nonsingular skew-symmetric form and hence $n - 2$ is even and $K_{x_1} \cap K_{x_2}$ is an orthogonal sum of hyperbolic planes, $K_{x_1} \cap K_{x_2} = P_1 \perp \cdots \perp P_r$, $2r = n - 2$, and $V = \langle x_1, x_2 \rangle \perp P_1 \perp \cdots \perp P_r$.

Theorem 3

Let $\Gamma:V \times V \to D$ be a Hermitian form. Then V is an orthogonal sum of lines, that is, there exists a basis $B = \{x_1, \cdots, x_n\}$ of V such that $m(\Gamma, B)$ is a diagonal matrix.

Proof

Note that if $\Gamma(x, x) = 0$ for all $x \in V$ then $\Gamma(x, y) = -\Gamma(y, x) = \alpha\Gamma(y, x)^\sigma$. If $\Gamma(y, x) = 0$ for all $x, y \in V$ then any basis will do. If not, then $\alpha = -1$, and σ is the identity map, which is a contradiction. Now the same induction argument as that of Theorem 1 proves the theorem.

DEFINITION 3 Let $\Gamma: V \times V \to D$ be a σ-bilinear form. A homomorphism $f: V \to V$ is called an *isometry* if $\Gamma(x, y) = \Gamma((x)f, (y)f)$ for all $x, y \in V$.

Proposition 1

Let $\Gamma: V \times V \to D$ be a σ-bilinear form.

(a) If $f, g: V \to V$ are isometries, then fg is an isometry.
(b) If Γ is nonsingular and f is an isometry, then f is a monomorphism. If V is finite dimensional, f is an isomorphism.

Proof

(a) $\Gamma(x, y) = \Gamma((x)f, (y)f) = \Gamma((x)fg, (y)fg)$ for all $x, y \in V$.
(b) If $(x)f = O$, then $\Gamma(x, y) = \Gamma((x)f, (y)f) = 0$ for all $y \in V \Rightarrow x = O$.

EXERCISES

1. If $\Gamma: V \times V \to D$ is a nonsingular σ-bilinear form, show that the set of all isometries is a subgroup of $GL(V)$.

2. Let $\dim V = 2$, $\Gamma: V \times V \to D$ a skew-symmetric form and $B = \{x_1, x_2\}$ a hyperbolic pair. Let $f: V \to V$ be a homomorphism and

$$m(f, B) = \begin{pmatrix} \alpha_{11} & \alpha_{12} \\ \alpha_{21} & \alpha_{22} \end{pmatrix}.$$

Show that f is an isometry $\Leftrightarrow \alpha_{11}\alpha_{22} - \alpha_{12}\alpha_{21} = 1$.

3. Let $V = \mathbb{R}^3$, $B = \{(1, 0, 0), (0, 1, 0), (0, 0, 1)\}$, and $\Gamma: V \times V \to \mathbb{R}$ the symmetric bilinear form such that

$$m(\Gamma, B) = \begin{pmatrix} 0 & 1 & 1 \\ 1 & 0 & 1 \\ 1 & 1 & 1 \end{pmatrix}.$$

Find a basis B' of $\mathbb{R}^3$ such that $m(\Gamma, B')$ is diagonal.

4. State and prove a generalization of Theorem 2 for skew-symmetric forms which are singular.

*5. Show that a matrix $A \in D_n$ is symmetric (that is, $A = A^T$) $\Leftrightarrow$ PAP^T is symmetric $\forall P \in D_n$. [4.9]

SECTION 3.4

QUADRATIC FORMS AND CONIC SECTIONS

Throughout this section let D be a field, $2^{-1} \in D$, ${}_DV$ a vector space, and $\Gamma: V \times V \to D$ a symmetric bilinear form. Associated with Γ is a function $Q: V \to D$ defined by $Q(x) = \Gamma(x, x)$. Q is called a *quadratic form.*

Proposition 1

(a) $Q(\alpha x) = \alpha^2 Q(x)$ for $\alpha \in D$, $x \in V$.
(b) $Q(x + y) + Q(x - y) = 2Q(x) + 2Q(y)$.
(c) $\Gamma(x, y) = 2^{-1}(Q(x + y) - Q(x) - Q(y))$.

Proof

(a) $Q(\alpha x) = \Gamma(\alpha x, \alpha x) = \alpha^2\Gamma(x, x) = \alpha^2 Q(x)$.

(b)
$$\begin{aligned} Q(x + y) + Q(x - y) &= \Gamma(x + y, x + y) + \Gamma(x - y, x - y). \\ &= \Gamma(x, x) + \Gamma(y, x) + \Gamma(x, y) + \Gamma(y, y) \\ &\quad + \Gamma(x, x) - \Gamma(y, x) - \Gamma(x, y) + \Gamma(y, y) \\ &= 2\Gamma(x, x) + 2\Gamma(y, y) = 2Q(x) + 2Q(y). \end{aligned}$$

(c)
$$\begin{aligned} \Gamma(x + y, x + y) = \Gamma(x, x) + 2\Gamma(x, y) + \Gamma(y, y) \Rightarrow \Gamma(x, y) \\ = 2^{-1}(Q(x + y) - Q(x) - Q(y)). \end{aligned}$$

As has been pointed out previously, we cannot use bilinear forms to define distance over fields such as $\mathbb{Z}_p$; however, for the purpose of conjecturing theorems and proofs it is an aid to intuition to think of $Q(x)$ as the square of the length of x (as it is if $D = \mathbb{R}$, $V = \mathbb{R}^2$, and Γ is the usual inner product). With this interpretation, part (b) of Proposition 1 is the assertion that the sum of the squares of the lengths of the diagonals of a parallelogram is equal to the sum of the squares of the lengths of the sides.

From the geometric point of view, an important interpretation of symmetric bilinear forms is in terms of certain polynomials. To avoid confusing indeterminates with vectors we shall use X's and Y's instead of x's and y's for the indeterminates of a polynomial. Let D be a field and $D[X, Y]$ the set of polynomials in two indeterminates with coefficients from D. Let $p(X, Y) \in D[X, Y]$ be the polynomial $p(X, Y) = 2XY$. If we take $B = \{(1, 0), (0, 1)\}$ as a basis for D^2 and $\Gamma: V \times V \to D$ to be the symmetric bilinear form such that

$$m(\Gamma, B) = \begin{pmatrix} 0 & 1 \\ 1 & 0 \end{pmatrix},$$

then Γ and p are related. The relationship is, to be precise, that if $\lambda \in D$, then $\{x = (\alpha, \beta) \in D^2 : \Gamma(x, x) = \lambda\}$ is precisely the set of $x = (\alpha, \beta)$ that satisfy the equation $p(X, Y) = \lambda$, since

$$\Gamma(x, x) = (\alpha\ \beta)\begin{pmatrix} 0 & 1 \\ 1 & 0 \end{pmatrix}\begin{pmatrix} \alpha \\ \beta \end{pmatrix} = (\beta\ \alpha)\begin{pmatrix} \alpha \\ \beta \end{pmatrix} = 2\alpha\beta.$$

This relationship can be extended to homogeneous polynomials of degree 2 as follows. Suppose $2^{-1} \in D$, $B = \{x_1, \cdots, x_n\}$ is a basis of ${}_DV$ and we have $p \in D[X_1, \cdots, X_n]$ of the form

$$p(X_1, \cdots, X_n) = \sum_{i,j} \alpha_{ij} X_i X_j, \; 1 \le i \le j \le n.$$

Let $\gamma_{ij} = \gamma_{ji} = 2^{-1}\alpha_{ij}$ and $\Gamma: V \times V \to D$ the bilinear form such that $m(\Gamma, B) = (\gamma_{ij})$. We sum over $i \le j$ because $X_iX_j = X_jX_i$ and if they appeared separately their coefficients would have to be combined to compute γ_{ij} and γ_{ji}.

Theorem 1

The map $p \to \Gamma$ described above is a 1-1 correspondence between symmetric bilinear forms on an n-dimensional vector space and homogeneous polynomials of degree 2 in n indeterminates.

Proof

The proof is left as an exercise for the reader.

This identification allows us to do some of the analytical geometry of conics and quadric surfaces in our context.

DEFINITION 1 Let ${}_DV$ be a vector space and $\Gamma: V \times V \to D$ a symmetric bilinear form. We let

$$G(\lambda) = \{x \in V : \Gamma(x, x) = \lambda\},$$
$$P_z(\lambda) = \{x \in V : \Gamma(x, z) = \lambda\}.$$

$G(\lambda)$ is called the *λ-graph* of Γ, and $P_z(\lambda)$ the *λ-pole* of Γ at z. Both, of course, depend on Γ; but, rather than indicate that in the notation, we shall rely on the context to make clear which Γ is being considered.

Proposition 2

$P_z(\lambda)$ is an affine subspace of V.

Proof

$x, y \in P_z(\lambda)$, $\alpha \in D \Rightarrow$

$$\begin{aligned}\Gamma(\alpha x + (1-\alpha)y, z) &= \alpha\Gamma(x, z) + (1-\alpha)\Gamma(y, z)\\ &= \alpha\lambda + (1-\alpha)\lambda = \lambda.\end{aligned}$$

Let us analyze in some detail the hyperbola $XY = 1$ over the field D, $2^{-1} \in D$. Letting $V = D^2$, $B = \{(1, 0), (1, 0)\}$, the appropriate bilinear form Γ has the matrix

$$m(\Gamma, B) = \begin{pmatrix} 0 & \frac{1}{2} \\ \frac{1}{2} & 0 \end{pmatrix}.$$

Then the hyperbola corresponds to $G(1)$. (See Figure 3.1.) $(1, 1) \in G(1)$ and

$$\begin{aligned}P_{(1,1)}(1) &= \left\{(\alpha, \beta) \in D^2 : (\alpha\ \beta)\begin{pmatrix} 0 & \frac{1}{2} \\ \frac{1}{2} & 0 \end{pmatrix} \cdot \begin{pmatrix} 1 \\ 1 \end{pmatrix} = 1\right\}\\ &= \{(\alpha, \beta) \in D^2 : \tfrac{1}{2}\beta + \tfrac{1}{2}\alpha = 1\}\\ &= (1, 1) + \langle(-1, 1)\rangle.\end{aligned}$$

Note that this line is, in $\mathbb{R}^2$, the tangent to the hyperbola at (1, 1). Indeed, you can check that if $z \in G(1)$, $P_z(1)$ is the tangent to the hyperbola at z. We have here, it would seem, a way to define algebraically tangents to quadric surfaces of various dimensions. The subject is made difficult by the fact that we sometimes get bad answers. For example, in $\mathbb{R}^3$ the tangent

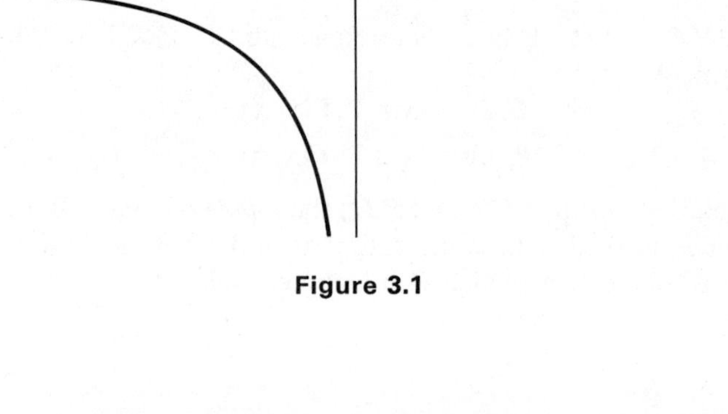

Figure 3.1

should be a plane, but the tangent, so defined, to $X^2 + Y^2 - Z^2 = 0$ at $(0, 0, 0)$ is $\mathbb{R}^3$. The situation will be reviewed in projective geometry where the algebraic characterizations are somewhat simpler.

To get back to the hyperbola, let us consider the isometries of Γ. Suppose $f: D^2 \to D^2$ is a homomorphism and

$$m(f, B) = \begin{pmatrix} \alpha_{11} & \alpha_{12} \\ \alpha_{21} & \alpha_{22} \end{pmatrix}.$$

Then f is an isometry

$$\begin{pmatrix} 0 & \frac{1}{2} \\ \frac{1}{2} & 0 \end{pmatrix} = \begin{pmatrix} \alpha_{11} & \alpha_{12} \\ \alpha_{21} & \alpha_{22} \end{pmatrix} \begin{pmatrix} 0 & \frac{1}{2} \\ \frac{1}{2} & 0 \end{pmatrix} \begin{pmatrix} \alpha_{11} & \alpha_{21} \\ \alpha_{12} & \alpha_{22} \end{pmatrix}$$

$$= \begin{pmatrix} \dfrac{\alpha_{12}}{2} & \dfrac{\alpha_{11}}{2} \\ \dfrac{\alpha_{22}}{2} & \dfrac{\alpha_{21}}{2} \end{pmatrix} \begin{pmatrix} \alpha_{11} & \alpha_{21} \\ \alpha_{12} & \alpha_{22} \end{pmatrix}$$

$$= \begin{pmatrix} \alpha_{11}\alpha_{12} & \dfrac{\alpha_{11}\alpha_{22} + \alpha_{12}\alpha_{21}}{2} \\ \dfrac{\alpha_{11}\alpha_{22} + \alpha_{12}\alpha_{21}}{2} & \alpha_{21}\alpha_{22} \end{pmatrix},$$

so $\alpha_{11}\alpha_{12} = \alpha_{21}\alpha_{22} = 0$ and $\alpha_{11}\alpha_{22} + \alpha_{12}\alpha_{21} = 1$. If $\alpha_{11} = 0$, then $\alpha_{12}\alpha_{21} = 1$ and $\alpha_{22} = 0$. If $\alpha_{11} \neq 0$, $\alpha_{11}\alpha_{22} = 1$ and $\alpha_{21} = 0$. The isometries of Γ are then precisely those homomorphisms whose matrices are of the form

$$\begin{pmatrix} \mu & 0 \\ 0 & \mu^{-1} \end{pmatrix} \quad \text{or} \quad \begin{pmatrix} 0 & \tau \\ \tau^{-1} & 0 \end{pmatrix}, \qquad \mu \neq 0, \quad \tau \neq 0.$$

The group of isometries of Γ acts *transitively* on $G(1)$, that is, if $x, y \in G(1)$ there exists an isometry f of Γ such that $(x)f = y$. This follows from the fact that if $x = (\alpha, \beta) \in G(1)$ then $\alpha\beta = 1$, that is, $\beta = \alpha^{-1}$ and so

$$x = (1, 1)\begin{pmatrix} \alpha & 0 \\ 0 & \beta \end{pmatrix},$$

that is, any point in $G(1)$ is an image of $(1, 1)$ and, since the inverse of an isometry is an isometry, $(1, 1)$ is the image of any point in $G(1)$.

The ellipse lends itself to a similar analysis, as you will see in the exercises. The parabola seems, at first glance, not to lend itself to this type of discussion because the polynomial $X^2 - Y$ is not homogeneous of degree 2. This can, however, be handled by considering the polynomial $X^2 - YZ$, the associated symmetric bilinear form, and the isometries of the bilinear form that map $\{(\alpha, \beta, \gamma) \in D^3 : \gamma = 1\}$ into itself (that is, restricting our attention to $Z = 1$).

EXERCISES

1. Prove Theorem 1.

2. Consider the ellipse $\mu X^2 + \nu Y^2 = 1$, $\mu > 0$, $\nu > 0$, in $\mathbb{R}^2$. Show that the group of isometries of the associated symmetric bilinear form is

$$\left\{ \begin{pmatrix} \pm\cos\theta & \pm\sqrt{\dfrac{\mu}{\nu}}\sin\theta \\ \pm\sqrt{\dfrac{\nu}{\mu}}\sin\theta & \pm\cos\theta \end{pmatrix} : \begin{array}{l} 0 \leq \theta \leq \pi/2, \text{ and exactly} \\ \text{three entries have the} \\ \text{same sign} \end{array} \right\}.$$

Does this group act transitively on the ellipse? If $z \in \mathbb{R}^2$, find $P_z(1)$. [*Hint*: Consider the cases (i) z outside the ellipse, (ii) z on the ellipse, (iii) z inside the ellipse.]

3. Consider the parabola $Y = X^2$ and, using the suggestions in the text, find the group of isometries. Does this group act transitively? If $z \in \mathbb{R}^2$ find $P_z(0)$.

4. Theorem 1, Section 3.3 said that given any symmetric bilinear form Γ there exists a basis B such that $m(\Gamma, B)$ is diagonal. Show that this is equivalent to the assertion that any homogeneous polynomial of degree 2 can be written as the sum of squares of linear combinations of the indeterminates. (In plane analytic geometry you phrased this by saying that there existed a rotation that would eliminate the XY term in an arbitrary second-degree equation.) Perform this "rotation" for $p(X, Y, Z) = 2XY + 2XZ + 2YZ$.

SECTION 3.5
THE INNER PRODUCT

DEFINITION 1 Let ${}_{\mathbb{R}}V$ be a vector space and $\Gamma: V \times V \to \mathbb{R}$ a symmetric bilinear form. Γ is called an *inner product* if $\Gamma(x, x) \geq 0$ for all $x \in V$, and $\Gamma(x, x) = 0 \Rightarrow x = O$. If Γ is an inner product we shall denote $\Gamma(x, y)$ by $x \cdot y$. A real vector space with an inner product is called an *inner product space*. If V is an inner product space, the length of $x \in V$, denoted $\|x\|$, is defined by $\|x\| = \sqrt{x \cdot x}$.

Proposition 1

Let V be an inner product space.

(a) $\|\alpha x\| = |\alpha| \, \|x\|$ for $\alpha \in \mathbb{R}$, $x \in V$. ($|\alpha|$ = absolute value of α.)
(b) Schwarz inequality: If $x, y \in V$ then $|x \cdot y| \leq \|x\| \cdot \|y\|$. Moreover, $|x \cdot y| = \|x\| \cdot \|y\| \Leftrightarrow \{x, y\}$ is linearly dependent.
(c) $\|x + y\| \leq \|x\| + \|y\|$ for all $x, y \in V$.

Proof

(a) $\|\alpha x\|^2 = (\alpha x) \cdot (\alpha x) = \alpha^2 (x \cdot x)$, so $\|\alpha x\| = |\alpha| \sqrt{x \cdot x} = |\alpha| \cdot \|x\|$.
(b) For $\alpha, \beta \in \mathbb{R}$ we have

$$\begin{aligned} 0 &\leq (\alpha x + \beta y) \cdot (\alpha x + \beta y) \\ &= \alpha^2 x \cdot x + 2\alpha\beta x \cdot y + \beta^2 y \cdot y; \end{aligned}$$

hence,

$$-2\alpha\beta x \cdot y \leq \alpha^2 \|x\|^2 + \beta^2 \|y\|^2.$$

If $x = O$ or $y = O$, then clearly $x \cdot y = \|x\| \cdot \|y\| = 0$, so assume $x \neq O$ and $y \neq O$. Then setting $\alpha = \|y\|$, $\beta = \|x\|$ yields $x \cdot y \geq -\|x\| \cdot \|y\|$ and setting $\alpha = -\|y\|$, $\beta = \|x\|$ yields $x \cdot y \leq \|x\| \cdot \|y\|$, so $|x \cdot y| \leq \|x\| \cdot \|y\|$. If x, y is linearly independent and $\alpha \neq 0$, $\beta \neq 0$, then $0 < (\alpha x + \beta y) \cdot (\alpha x + \beta y)$ and $|x \cdot y| < \|x\| \cdot \|y\|$. If $x \neq O$, $y \neq O$ are linearly dependent, $x = \alpha y$ for some $\alpha \in \mathbb{R}$, whence

$$|x \cdot y| = |\alpha y \cdot y| = |\alpha| \cdot |y \cdot y| = |\alpha| \cdot \|y\| \cdot \|y\| = \|x\| \cdot \|y\|.$$

(c) If $x, y \in V$, then

$$\begin{aligned} \|x + y\|^2 &= (x + y) \cdot (x + y) \\ &= x \cdot x + 2x \cdot y + y \cdot y \\ &\leq \|x\|^2 + 2|x \cdot y| + \|y\|^2 \\ &\leq \|x\|^2 + 2\|x\| \cdot \|y\| + \|y\|^2 \\ &\leq (\|x\| + \|y\|)^2 \\ &\Rightarrow \|x + y\| \leq \|x\| + \|y\|. \end{aligned}$$

DEFINITION 2 If ${}_{\mathbb{R}}V$ is an inner product space and $x, y \in V$ the *distance from x to y*, denoted $d(x, y)$, is defined by $d(x, y) = \|x - y\|$.

Proposition 2

If V is an inner product space the distance function satisfies the following properties:

(a) $d(x, y) \geq 0$;
(b) $d(x, y) = 0 \Rightarrow x = y$;
(c) $d(x, y) = d(y, x)$;
(d) $d(x, y) \leq d(x, z) + d(z, y)$.

Proof

The proof is left as an exercise for the reader.

Let $x, y \in V, y \neq O$. Then there exists a unique $\alpha \in \mathbb{R}$ such that $(x - \alpha y) \cdot y = 0$. To be precise, $\alpha = (1/\|y\|^2)x \cdot y$. The observation that

$$\frac{\|\alpha y\|}{\|x\|} = \frac{|\alpha| \, \|y\|}{\|x\|} = \frac{|x \cdot y| \cdot \|y\|}{\|y\|^2 \, \|x\|} = \frac{|x \cdot y|}{\|x\| \cdot \|y\|}$$

motivates the following definition. (See Figure 3.2.)

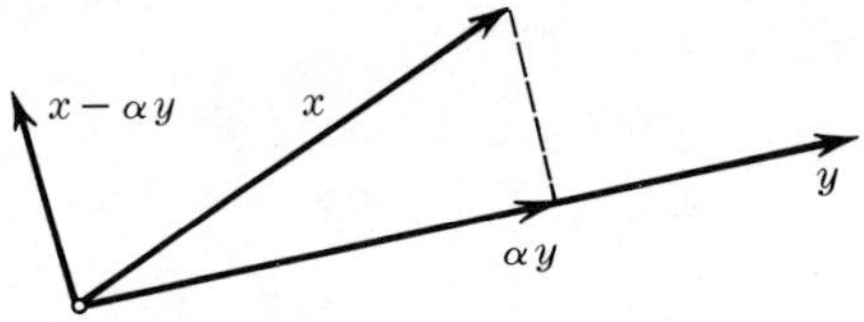

Figure 3.2

DEFINITION 3 Let ${}_{\mathbb{R}}V$ be an inner product space and x, y nonzero vectors. The angle $\theta(x, y)$ between x and y is defined to be the unique number, $0 \leq \theta(x, y) \leq \pi$, such that

$$\cos(\theta(x, y)) = \frac{x \cdot y}{\|x\| \, \|y\|}.$$

Proposition 3 (Law of Cosines)

If ${}_{\mathbb{R}}V$ is an inner product space, $x, y \in V$, $x \neq O$, $y \neq O$, then

$$\|x - y\|^2 = \|x\|^2 + \|y\|^2 - 2\,\|x\| \cdot \|y\| \cos \theta(x, y).$$

Proof

$$\begin{aligned} \|x - y\|^2 &= (x - y) \cdot (x - y) = x \cdot x + y \cdot y - 2x \cdot y \\ &= \|x\|^2 + \|y\|^2 - 2\,\|x\| \cdot \|y\| \cos(\theta(x, y)). \end{aligned}$$

Since an inner product is a symmetric bilinear form, we know (Theorem 1, Section 3.3) that there exists a basis $B = \{x_1, \cdots, x_n\}$ of U such that if $i \neq j$, $x_i \cdot x_j = 0$. Since $x_i \cdot x_i = \alpha_i > 0$ we can set $x_i' = (1/\sqrt{\alpha_i})x_i$; let $B' = \{x_1', \cdots, x_n'\}$; we have a basis such that $x_i' \cdot x_j' = 0$ for $i \neq j$ and $x_i' \cdot x_i' = 1$ for $i = 1, \cdots, n$. A basis with this property is said to be *orthonormal*. Although the existence of an orthonormal basis is, as we have seen, a triviality, the following is a somewhat deeper result.

Theorem 1 (Gram-Schmidt)

Let ${}_{\mathbb{R}}V$ be an inner product space and $\{x_1, \cdots, x_n\}$ a basis of V. Then there exists an orthonormal basis $\{z_1, \cdots, z_n\}$ of V such that $\langle x_1, \cdots, x_k \rangle = \langle z_1, \cdots, z_k \rangle$ for $1 \leq k \leq n$.

Proof

The proof is by induction. $x_1 \cdot x_1 = \alpha_1 > 0$. Set $z_1 = (1/\sqrt{\alpha_1})x_1$ so $z_1 \cdot z_1 = 1$. Suppose we have $z_1, \cdots, z_j$ linearly independent such that $z_i \cdot z_i = 1$, $1 \leq i \leq j$, $z_i \cdot z_k = 0$ for $i \neq k$ and $\langle z_1, \cdots, z_i \rangle = \langle x_1, \cdots, x_i \rangle$, $1 \leq i \leq j$. Set $\beta_i = x_{j+1} \cdot z_i$, $1 \leq i \leq j$, and $w = x_{j+1} - \beta_1 z_1 - \cdots - \beta_j z_j$. Then $w \cdot z_i = x_{j+1} \cdot z_i - \beta_i z_i \cdot z_i = \beta_i - \beta_i = 0$, $1 \leq i \leq j$; and $x_{j+1} \notin \langle x_1, \cdots, x_j \rangle = \langle z_1, \cdots, z_j \rangle$ so $w \neq O$. Let $z_{j+1} = (1/\|w\|)w$. Then $z_{j+1} \cdot z_{j+1} = 1$, and $z_{j+1} \cdot z_i = 0$ for $i \leq j$. By definition $z_{j+1} \in \langle x_{j+1} \rangle + \langle z_1, \cdots, z_j \rangle = \langle x_{j+1} \rangle + \langle x_1, \cdots, x_j \rangle = \langle x_1, \cdots, x_{j+1} \rangle$, and so $\langle z_1, \cdots, z_{j+1} \rangle \subseteq \langle x_1, \cdots, x_{j+1} \rangle$. Similarly, $\langle x_1, \cdots, x_{j+1} \rangle \subseteq \langle z_1, \cdots, z_{j+1} \rangle$, and the theorem is proven.

Corollary 1

If V is a finite dimensional inner product space and $U < V$ then $U \oplus U^{\perp} = V$.

Proof

Recall that $U^{\perp} = \{y \in V : y \cdot u = 0 \text{ for all } u \in U\}$. $x \in U \cap U^{\perp} \Rightarrow x \cdot x = 0 \Rightarrow x = O$, so $U + U^{\perp} = U \oplus U^{\perp}$. Using Theorem 1 we can extend an orthonormal basis $\{x_1, \cdots, x_k\}$ of U to an orthonormal basis $\{x_1, \cdots, x_k, x_{k+1}, \cdots, x_n\}$ of V. If $u \in U$ then $u = \alpha_1 x_1 + \cdots + \alpha_k x_k$, so $u \cdot x_j = \alpha_1 x_1 \cdot x_j + \cdots + \alpha_k x_k \cdot x_j = 0$ for $j > k$, and $x_j \in U^{\perp}$ so $\dim(U^{\perp}) \geq \dim(V) - \dim(U) \Rightarrow U \oplus U^{\perp} = V$.

Corollary 2

If $\{x_1, \cdots, x_k\}$ is orthonormal, that is $x_i \cdot x_j = \begin{cases} 1 & i = j \\ 0 & i \neq j \end{cases}$, then it is linearly independent and if $x \in \langle x_1, \cdots, x_k \rangle$, $x = (x \cdot x_1)x_1 + \cdots + (x \cdot x_n)x_n$.

Proof

If $O = \alpha_1 x_1 + \cdots + \alpha_k x_k$, then $0 = O \cdot x_i = (\alpha_1 x_1 + \cdots + \alpha_k x_k) \cdot x_i = \alpha_i$, $i = 1, \cdots, k$, and the linear independence is proven. If $x \in \langle x_1, \cdots, x_k \rangle$, then $x = \alpha_1 x_1 + \cdots + \alpha_k x_k$ and $x \cdot x_i = (\alpha_1 x_1 + \cdots + \alpha_k x_k) \cdot x_i = \alpha_i(x_i \cdot x_i) = \alpha_i$.

From this we can deduce a result which, in its obtaining algebraic information from a geometric assumption, has some of the flavor of Theorem 1 of Chapter 2, Section 2.6, which showed that 1-1 maps preserving affine collinearity are semilinear transformations.

Theorem 2

Let ${}_{\mathbb{R}}V$ be a finite dimensional inner product space and $f:V \to V$ a map such that $(x)f \cdot (y)f = x \cdot y$ for all $x, y \in V$. Then f is an isomorphism.

Proof

It should be noted that we have not assumed f to be a homomorphism. Let $\{x_1, \cdots, x_n\}$ be an orthonormal basis of V. Then $\{(x_1)f, \cdots, (x_n)f\}$ is an orthonormal basis of V also. Moreover, if $x \in V$, then $x = \alpha_1 x_1 + \cdots + \alpha_n x_n$, and, using Corollary 2,

$$\alpha_i = x \cdot x_i = (x)f \cdot (x_i)f \Rightarrow (x)f = \alpha_1(x_1)f + \cdots + \alpha_n(x_n)f,$$

and f is an isomorphism.

EXERCISES

1. If $\|x\| = \|y\|$ prove that $(x + y) \cdot (x - y) = 0$. (This says that the diagonals of a rhombus are perpendicular.)

2. (Bessel's inequality.) Prove: If ${}_{\mathbb{R}}V$ is an inner product space, $\{x_1, \cdots, x_n\}$ is an orthonormal basis, and $x \in V$, then for $m \leq n$, $\sum_{i=1}^{m} |x \cdot x_i|^2 \leq \|x\|^2$.

3. (Parseval's identity.) Prove: If V is an inner product space and $\{x_1, \cdots, x_n\}$ is an orthonormal basis, then $x \cdot y = \sum_{i=1}^{n} (x \cdot x_i)(y \cdot x_i)$.

*4. If ${}_{\mathbb{R}}V$ is an inner product space and $f:V \to V$ is a homomorphism that maps orthonormal bases to orthonormal bases, show that f is an isometry. In particular show that f preserves distance and angle between vectors. [3.6]

5. Let ${}_{\mathbb{C}}V$ be a complex vector space, let σ be conjugation, and let $\Gamma: V \times V \to \mathbb{C}$ be a Hermitian form. We know that $\Gamma(x, x) \in \mathbb{R}$ for all $x \in V$. (Why?) If in addition $\Gamma(x, x) \geq 0$ for all $x \in V$ and $\Gamma(x, x) = 0 \Rightarrow x = 0$, then Γ is called *unitary*. Develop a theory of unitary forms analogous to what was done in this section for inner products.

*6. Suppose ${}_{\mathbb{R}}V$ is an inner product space, with orthonormal basis $B = \{x_1, \cdots, x_n\}$. Let $\hat{V}$ be the complexification of V (see Chapter 1, Section 1.8, Exercise 8 for the definitions) and $\hat{B}$ the basis of $\hat{V}$ corresponding to B.

(a) Show that the form defined by

$$(x + yi) \cdot (x' + y'i) = (x \cdot x' + y \cdot y') + (y \cdot x' - x \cdot y')i$$

is a unitary form on $\hat{V}$.

(b) Show that $\hat{B}$ is an orthonormal basis for $\hat{V}$ relative to this induced unitary form. [3.8]

7. Let $(\alpha_1, \alpha_2) = u$ and $(\beta_1, \beta_2) = v$ be vectors in $\mathbb{R}^2$. Show that the area of the parallelogram $(O, u, u + v, v)$ is $|\alpha_1\beta_2 - \alpha_2\beta_1|$.

8. Let $B = \{x_1, \cdots, x_n\}$ be an orthonormal basis of ${}_{\mathbb{R}}V$ and $f: V \to V$ an isomorphism. Show that $(x)f \cdot (y)f = x \cdot y, \forall x, \ y \in V \Leftrightarrow m(f, B)^{-1} = m(f, B)^T$.

SECTION 3.6
DETERMINANTS

Those of you who have had some previous contact with determinants will notice that our approach is substantially different from the usual introduction to determinants. We make no claim that our approach is necessarily easier, but it does seem to have certain advantages. First, it acknowledges our emphasis on homomorphisms rather than matrices. Second, it has a firm geometric relationship with notions such as length, area, volume, and their generalizations.

For each field D and each positive integer n, we shall define a function $d: D_n \to D$ called the *determinant*. Equivalently, as we shall see, d may be viewed as a function from $\text{Hom}_D(V, V)$ into D, where ${}_DV$ has dimension n, the connection being that each homomorphism f corresponds to an $n \times n$ matrix relative to a fixed basis. Among other properties, d will have value 0 on matrices whose rows are linearly dependent (equivalently on homomorphisms with nonzero kernels), and d will be multiplicative: $d(AB) = d(A)d(B)$ (equivalently, $d(f \cdot g) = d(f) \cdot d(g)$).

In order to prove that such a function exists, we begin by considering a particular function on $\mathbb{R}^2$ having a nice geometric interpretation.

Exercise 6, Section 3.5 showed that if $u = (\alpha_1, \alpha_2)$ and $v = (\beta_1, \beta_2)$ are vectors in $\mathbb{R}^2$, then the area of the parallelogram $(O, u, v, u + v)$ is given by $|\alpha_1\beta_2 - \alpha_2\beta_1|$. We can view this relationship as a function Γ of $\mathbb{R}_2$ into $\mathbb{R}$, where

$$\Gamma\begin{pmatrix} \alpha_1 & \alpha_2 \\ \beta_1 & \beta_2 \end{pmatrix} = |\alpha_1\beta_2 - \alpha_2\beta_1|.$$

It is a simple matter to show that Γ is multiplicative in the sense that $\Gamma(AB) = \Gamma(A) \cdot \Gamma(B)$, for $A, B \in \mathbb{R}_2$, and that if u and v are linearly dependent, then

$$\Gamma\begin{pmatrix} \alpha_1 & \alpha_2 \\ \beta_1 & \beta_2 \end{pmatrix} = 0.$$

In fact, this function is very close to what we want when $n = 2$ and $D = \mathbb{R}$. However, in an arbitrary field, "absolute value" has no meaning, so that this definition would not generalize properly.

A small amount of hindsight would suggest that we consider the function

$$\Gamma\begin{pmatrix} \alpha_1 & \alpha_2 \\ \beta_1 & \beta_2 \end{pmatrix} = \alpha_1\beta_2 - \alpha_2\beta_1,$$

that is, forget the absolute value. If we do so, we lose, it would seem, some of the geometric flavor of the function; on the other hand, we gain more than enough to compensate for that.

Think of Γ now as a function from $\mathbb{R}^2 \times \mathbb{R}^2$ into $\mathbb{R}$ defined by $\Gamma((\alpha_1, \alpha_2), (\beta_1, \beta_2)) = \alpha_1\beta_2 - \alpha_2\beta_1$. Up to sign Γ gives the area of the parallelogram $(O, u, v, u + v)$. As Figure 3.3 illustrates, the area of $(O, \alpha u, v, \alpha u + v)$ should be $|\alpha|$ times the area of $(O, u, v, u + v)$.

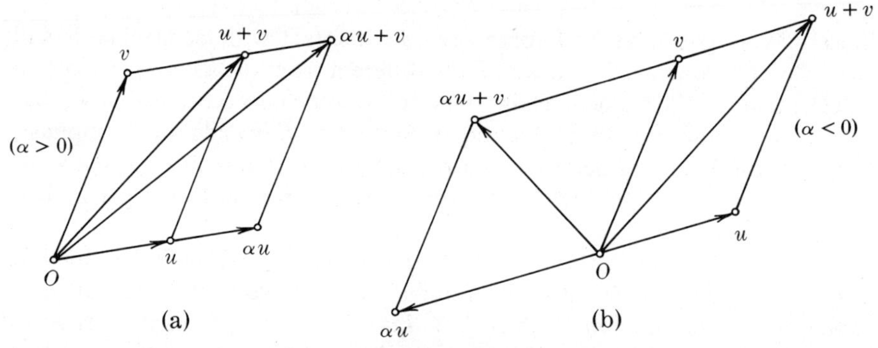

Figure 3.3

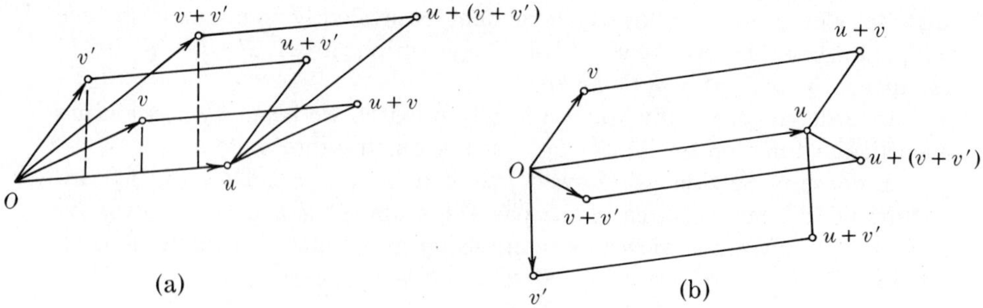

Figure 3.4

In terms of Γ, though, $\Gamma(\alpha u, v) = \alpha\Gamma(u, v)$, and $\Gamma(u, \alpha v) = \alpha\Gamma(u, v)$, by direct computation, regardless of the sign of α.

As Figure 3.4 illustrates, the area of $(O, u, v + v', u + v + v')$ should be the sum of the areas of $(O, u, v, u + v)$ and $(O, u, v', u + v')$, *provided* v and v' are both "on the same side of u"; otherwise additivity does not hold. But in terms of Γ, direct computation gives

$$\Gamma(u, v + v') = \Gamma(u, v) + \Gamma(u, v'),$$

and

$$\Gamma(u + u', v) = \Gamma(u, v) + \Gamma(u', v).$$

In short, Γ is a bilinear form on $\mathbb{R}^2$ with the additional property that $\Gamma(u, u) = 0$.

In $\mathbb{R}^3$ any three vectors, u, v, w determine a parallelepiped $(O, u, v, w, u + v, u + w, v + w, u + v + w)$ whose volume, up to sign, is given by a function $\Theta(u, v, w)$ that is linear in each variable and has value 0 if $u = v$, $u = w$, or $v = w$. Without being explicit, it is clear from the foregoing example that such properties should hold.

Now we let D be an arbitrary field, and ${}_DV$ be any space of dimension n. We denote by V^k the Cartesian product of k copies of V, for $k = 1, 2, \cdots, n$.

DEFINITION 1 Let $1 \leq k \leq n$. A function $\Theta: V^k \to D$ such that:

(1) $$\Theta(x_1, \cdots, \alpha x_i + \beta x_i', \cdots, x_k) = \alpha\Theta(x_1, \cdots, x_i, \cdots, x_k) + \beta\Theta(x_1, \cdots, x_i', \cdots, x_k),$$

$$\forall i = 1, 2, \cdots, k, \ \forall \alpha, \beta \in D, \ \forall x_1, \cdots, x_i, x_i', \cdots, x_k \in V,$$

and

(2) $\Theta(x_1, \cdots, x_k) = 0$ if $x_i = x_j$ for some $i \neq j$, is called an *alternating k-linear form* on V. Any Θ satisfying (1) is called *k-linear*.

If $k = 1$, condition (2) is vacuously satisfied so any element of the dual space is an alternating 1-linear form. A skew-symmetric bilinear form is an alternating 2-linear form.

We shall make use of the notion of a k-linear form with $k \leq n$ for induction purposes in showing the existence of a determinant function. Intuitively, an alternating k-linear form on an n-dimensional space should be viewed as a function that gives a k-dimensional "content" of a k-dimensional parallelepiped. Such a situation arises when one wants to assign an "area" to a parallelogram (2-dimensional) or a "length" to a vector (1-dimensional) in $\mathbb{R}^3$. We offer a sketch of one method of solving the "area" problem. A more rigorous development of this problem is given in the exercises.

Suppose M is a plane in $\mathbb{R}^3$ passing through O. Choose an orthonormal basis $\{x_1, x_2, x_3\}$ of $\mathbb{R}^3$ such that $M = \langle x_1, x_2 \rangle$. The linear map $f: \mathbb{R}^3 \to \mathbb{R}^3$

given by $(x_1)f = (1, 0, 0)$, $(x_2)f = (0, 1, 0)$, and $(x_3)f = (0, 0, 1)$ is a rigid motion of $\mathbb{R}^3$, preserving distance, area, and volume. The plane M is mapped onto the xy-plane. For two vectors $u' = (\alpha_1, \alpha_2, 0)$ and $v' = (\beta_1, \beta_2, 0)$ in the xy-plane we have the (signed) area $\Theta(u', v') = \alpha_1\beta_2 - \alpha_2\beta_1$ of the parallelogram $(O, u', u' + v', v')$. Now it is easy to define a function Θ' on pairs of vectors (u, v) in M that assigns a (signed) area of the parallelogram $(O, u, u + v, v)$, namely,

$$\Theta'(u, v) = \Theta((u)f, (v)f).$$

It should be pointed out that the function Θ' is defined only on M. But by extending Θ in the obvious way, Θ' extends to any pair of vectors in $\mathbb{R}^3$. Even so, Θ' does not give the correct value for area of any parallelogram not in M. It does, however, assign a 2-dimensional content (area) to certain 2-dimensional parallelepipeds (parallelograms in M) in a 3-dimensional space ($\mathbb{R}^3$). We proceed now with the development.

Proposition 1

Let $\Theta: V^k \to D$ be an alternating k-linear form. For $i \neq j$,

$$\Theta(x_1, \cdots, x_i, \cdots, x_j, \cdots, x_k) = -\Theta(x_1, \cdots, x_j, \cdots, x_i, \cdots, x_k).$$

Proof

The proof is by condition (2) of Definition 1,

$$\begin{aligned} 0 &= \Theta(x_1, \cdots, x_i + x_j, \cdots, x_j + x_i, \cdots, x_k) \\ &= \Theta(x_1, \cdots, x_i, \cdots, x_j, \cdots, x_k) + \Theta(x_1, \cdots, x_i, \cdots, x_i, \cdots, x_k) \\ &\quad + \Theta(x_1, \cdots, x_j, \cdots, x_j, \cdots, x_k) + \Theta(x_1, \cdots, x_j, \cdots, x_i, \cdots, x_k), \end{aligned}$$

and the second and third summands are 0.

Proposition 2

Let $\Theta: V^k \to D$ be an alternating k-linear form. If $x_1, \cdots, x_k$ are linearly dependent, then $\Theta(x_1, \cdots, x_k) = 0$.

Proof

If $\{x_1, \cdots, x_k\}$ are linearly dependent, then $x_i = \alpha_1 x_1 + \cdots + \alpha_{i-1}x_{i-1}$, for some $i \leq k$, and some $\alpha_1, \cdots, \alpha_{i-1} \in D$. Then

$$\begin{aligned} \Theta(x_1, \cdots, x_k) &= \Theta(x_1, \cdots, \alpha_1 x_1 + \cdots + \alpha_{i-1}x_{i-1}, \cdots, x_k) \\ &= \alpha_1 \Theta(x_1, \cdots, x_1, \cdots, x_k) + \cdots \\ &\qquad + \alpha_{i-1}\Theta(x_1, \cdots, x_{i-1}, x_{i-1}, \cdots, x_k) = 0. \end{aligned}$$

The next theorem is a simple extension of the following idea: Suppose Γ is a 2-linear alternating form on V, and Λ is a 1-linear form on V. Define a 3-linear form $\Theta: V^3 \to D$ by

$$\Theta(x_1, x_2, x_3) = \Lambda(x_1)\Gamma(x_2, x_3) - \Lambda(x_2)\Gamma(x_1, x_3) + \Lambda(x_3)\Gamma(x_1, x_2).$$

Linearity in each variable follows from the fact that each term $\Gamma(\_\_, \_\_) \cdot \Lambda(\_\_)$ is linear in each variable. (See Exercise 1.) To check that Θ is alternating, note that if $x_1 = x_2$ then the third term is 0 and the first two are negatives of each other by Proposition 1. If $x_1 = x_3$, then the second term is 0, and the remaining two cancel. If $x_2 = x_3$, then the first term is 0, and the second and third cancel.

Theorem 1

If $1 \le k \le n$, there exists an alternating k-linear form which is not identically 0.

Proof

The proof is by induction. Let $k = 1$. Choose $x \in V$, $x \neq 0$, so that $V = \langle x \rangle \oplus U$. If $y \in V$, then $y = \alpha x + u$. Define $\Theta(y) = \alpha$. Then Θ is certainly linear, and condition (2) of Definition 1 is vacuously satisfied.

Suppose we have an alternating $k - 1$ linear form Λ, with $\Lambda(x_1, \cdots, x_{k-1}) \neq 0$. By Proposition 2, $x_1, \cdots, x_{k-1}$ are linearly independent and so can be extended to a basis $\{x_1, \cdots, x_{k-1}, x_k, \cdots, x_n\}$ of V. By Theorem 1 of Chapter 1, Section 1.7, there is a unique homomorphism $\psi: V \to D$ such that $(x_i)\psi = \begin{Bmatrix} 0 & i \neq k \\ 1 & i = k \end{Bmatrix}$. We define $\Theta: V^k \to D$ by

$$\Theta(y_1, \cdots, y_k) = \sum_{i=1}^{k} (-1)^{i+1} (y_i)\psi\Lambda(y_1, \cdots, \hat{y}_i, \cdots, y_k),$$

where

$$(y_1, \cdots, \hat{y}_i, \cdots, y_k) = (y_1, \cdots, y_{i-1}, y_{i+1}, \cdots, y_k).$$

Θ is linear in each variable because each summand is. Suppose $y_p = y_j$, for $1 \le p < j \le k$. Then

$$\begin{aligned}\Theta(y_1, \cdots, y_k) &= \sum_{i=1}^{k} (-1)^{i+1} (y_i)\psi\Lambda(y_1, \cdots, \hat{y}_i, \cdots, y_k) \\ &= (-1)^{p+1}(y_p)\psi\Lambda(y_1, \cdots, \hat{y}_p, \cdots, y_k) \\ &\quad + (-1)^{j+1}(y_j)\psi\Lambda(y_1, \cdots, \hat{y}_j, \cdots, y_k),\end{aligned}$$

for all other terms in the summation have Λ evaluating a $(k-1)$-tuple, containing both y_p and y_j, and hence vanishing. Observing that the factor $\Lambda(y_1, \cdots, \hat{y}_p, \cdots, y_j, \cdots, y_k)$ becomes $\Lambda(y_1, \cdots, y_p, \cdots, \hat{y}_j, \cdots, y_k)$ upon interchanging y_j with its predecessor $j-p-1$ times and that each such interchange gives us a new factor of -1, we have

$$\begin{aligned}\Theta(y_1, \cdots, y_k) &= (-1)^{p+1}(y_p)\psi(-1)^{j-p-1}\Lambda(y_1, \cdots, \hat{y}_j, \cdots, y_k) \\ &\quad + (-1)^{j+1}(y_p)\psi\Lambda(y_1, \cdots, \hat{y}_j, \cdots, y_k) \\ &= 0.\end{aligned}$$

DEFINITION 2 We call a k-linear form Θ on V *nontrivial* if $\Theta(y_1, \cdots, y_k) \neq 0$ for some $y_1, \cdots, y_k \in V$.

Theorem 2

(a) Let Θ be a nontrivial alternating n-linear form on V, where $\dim V = n$. If $\{x_1, \cdots, x_n\}$ is linearly independent, then $\Theta(x_1, \cdots, x_n) \neq 0$.

(b) Let Θ, Λ be nontrivial alternating n-linear forms on V where $\dim V = n$. Then there is a unique $\lambda \in D$ such that $\Lambda = \lambda\Theta$, that is,

$$\Lambda(x_1, \cdots, x_n) = \lambda\Theta(x_1, \cdots, x_n), \forall x_1, \cdots, x_n \in D.$$

Proof

(a) Since Θ is nontrivial, $\exists y_1, \cdots, y_n \in V$ such that $\Theta(y_1, \cdots, y_n) \neq 0$. Put $y_1 = \sum_{j=1}^{n} \alpha_{ij} x_j$. Then

$$\Theta(y_1, \cdots, y_n) = \Theta\left(\sum_{j=1}^{n} \alpha_{1j} x_j, \cdots, \sum_{j=1}^{n} \alpha_{nj} x_j\right).$$

By repeated use of linearity we expand $\Theta(y_1, \cdots, y_n)$ into a sum of terms of the form $\beta \cdot \Theta(x_{i_1}, \cdots, x_{i_n})$. Each of these terms is either 0 (if two or more of the x_{i_j}'s are equal) or $\pm\beta\Theta(x_1, \cdots, x_n)$ by repeated use of Proposition 1. Hence $\Theta(y_1, \cdots, y_n) = \mu \cdot \Theta(x_1, \cdots, x_n)$, for some $\mu \in D$. Consequently, $\Theta(x_1, \cdots, x_n) \neq 0$.

(b) Choose $\{x_1, \cdots, x_n\}$ to be a basis of V. Then by part (a), $\Theta(x_1, \cdots, x_n) \neq 0$, and $\Lambda(x_1, \cdots, x_n) \neq 0$. We put $\Lambda(x_1, \cdots, x_n) = \lambda \cdot \Theta(x_1, \cdots, x_n)$, for some $\lambda \in D$. Let $\{y_1, \cdots, y_n\} \subseteq V$. By part (a), $\exists \mu \in D$ such that

$$\Theta(y_1, \cdots, y_n) = \mu \cdot \Theta(x_1, \cdots, x_n).$$

In fact, in the expansion of $\Theta(y_1, \cdots, y_n)$ mentioned in part (a), if we replace Θ by Λ we arrive at

$$\Lambda(y_1, \cdots, y_n) = \mu \cdot \Lambda(x_1, \cdots, x_n),$$

for the *same* $\mu \in D$, since the expansion process used only the fact that Θ was an alternating n-linear form, and hence the value of μ depends only on the coefficients $\{\alpha_{ij}\}$. Thus we have

$$\begin{aligned}\Lambda(y_1, \cdots, y_n) &= \mu\Lambda(x_1, \cdots, x_n) = \mu \cdot \lambda \cdot \Theta(x_1, \cdots, x_n) \\ &= \lambda \cdot \mu \cdot \Theta(x_1, \cdots, x_n) = \lambda \cdot \Theta(y_1, \cdots, y_n).\end{aligned}$$

Theorem 3

Let $f:V \to V$ be a homomorphism, with $\dim V = n$. Then there exists a unique element of D, denoted $d(f)$, such that if Θ is any nontrivial alternating n-linear form on V, we have

$$\Theta((x_1)f, \cdots, (x_n)f) = d(f) \cdot \Theta(x_1, \cdots, x_n), \forall \{x_1, \cdots, x_n\} \subseteq V.$$

Proof

If $\operatorname{Ker}(f) \neq O$, then $\{(x_1)f, \cdots, (x_n)f\}$ is a linearly dependent set, for all $\{x_1, \cdots, x_n\} \subseteq V$ so that $\Theta((x_1)f, \cdots, (x_n)f) = 0$, for any choice of Θ. We put $d(f) = 0$.

If f is a monomorphism (and hence an isomorphism on V), then for any nontrivial Θ, the map $\Theta':V^n \to D$ defined by $\Theta'(x_1, \cdots, x_n) = \Theta((x_1)f, \cdots, (x_n)f)$ is a nontrivial alternating n-linear form. (Check this!) Hence by Theorem 2, there exists a unique $d(f) \in D$ such that

$$\Theta((x_1)f, \cdots, (x_n)f) = d(f)\Theta(x_1, \cdots, x_n), \forall x_1, \cdots, x_n \in V.$$

Moreover, $d(f)$ is independent of the choice of Θ, for if Λ is a second nontrivial form, $\Lambda = \lambda\Theta$ for some $\lambda \in D$, so that

$$\begin{aligned}\Lambda((x_1)f, \cdots, (x_n)f) &= \lambda\Theta((x_1)f, \cdots, (x_n)f) = \lambda \cdot d(f) \cdot \Theta(x_1, \cdots, x_n)\\ &= d(f) \cdot \Lambda(x_1, \cdots, x_n).\end{aligned}$$

DEFINITION 3 The value $d(f)$ is called the *determinant of f*.

Theorem 4

Let f, g be homomorphisms from V into V. Then $d(f \cdot g) = d(f) \cdot d(g)$.

Proof

Let Θ be a nontrivial alternating n-linear form on V.

$$\begin{aligned}d(f \cdot g)\Theta(x_1, \cdots, x_n) &= \Theta((x_1)fg, \cdots, (x_n)fg)\\ &= d(g)\Theta((x_1)f, \cdots, (x_n)f)\\ &= d(g)d(f)\Theta(x_1, \cdots, x_n), \forall \{x_1, \cdots, x_n\} \subseteq V.\end{aligned}$$

In particular, if $\{x_1, \cdots, x_n\}$ is a basis, then $\Theta(x_1, \cdots, x_n) \neq 0$, so that $d(f \cdot g) = d(g)d(f) = d(f)d(g)$.

Corollary

If 1 is the identity map on V, then $d(1) = 1$, and if f is an isomorphism, $d(f^{-1}) = d(f)^{-1}$, and, consequently $d(f) = 0 \Leftrightarrow \operatorname{Ker}(f) \neq O$.

The more traditional definition of determinants is given in terms of matrices; we proceed to make contact with that development. We let $(\alpha_{ij}) \in D_n$. Let ${}_DV$ have dimension n, and $B = \{x_1, \cdots, x_n\}$ be a basis of V. Define $d((\alpha_{ij})) = d(f)$, where f is the (unique) homomorphism such that $m(f, B) = (\alpha_{ij})$. It would appear that $d((\alpha_{ij}))$ so defined would depend on the choice of the basis. However, let $B' = \{x'_1, \cdots, x'_n\}$ be a second basis of V, and suppose $m(g, B') = (\alpha_{ij})$. We let $h:V \to V$ be the isomorphism defined by $(x_i)h = x'_i$. Then for each i,

$$(x_i)hgh^{-1} = (x'_i)gh^{-1} = \Big(\sum_j \alpha_{ij} x'_j\Big)h^{-1} = \sum_j \alpha_{ij} x_i = (x_i)f.$$

Then $hgh^{-1} = f$, so that

$$d(f) = d(hgh^{-1}) = d(h)d(g)d(h^{-1}) = d(g).$$

DEFINITION 4 The value $d((\alpha_{ij}))$ is called the *determinant of* (α_{ij}).

Actually, there is a more subtle point. The above definition of $d((\alpha_{ij}))$ seems to depend on the choice of the vector space V of dimension n. The next few results, however, show that this is not true, and d is in fact well defined. We suppose that some space V is fixed in advance and that d is defined on D_n as above.

Proposition 3

Let $\{x_1, \cdots, x_n\}$ be a basis of V, and let $f, g, h:V \to V$ be homomorphisms such that $(x_i)f = (x_i)g = (x_i)h$ for $i \neq k$, and $\alpha(x_k)f + \beta(x_k)g = (x_k)h$. Then $d(h) = \alpha d(f) + \beta d(g)$.

Proof

Let Θ be any nontrivial alternating n-linear form on V. Then

$$\begin{aligned} d(h)\Theta(x_1, \cdots, x_n) &= \Theta((x_1)h, \cdots, (x_k)h, \cdots, (x_n)h) \\ &= \Theta((x_1)h, \cdots, \alpha(x_k)f + \beta(x_k)g, \cdots, (x_n)h) \\ &= \alpha\Theta((x_1)h, \cdots, (x_k)f, \cdots, (x_n)h) + \beta \cdot \Theta((x_1)h + \cdots \\ &\qquad + (x_k)g, \cdots, (x_n)h) \\ &= \alpha d(f)\Theta(x_1, \cdots, x_n) + \beta d(g) \cdot \Theta(x_1, \cdots, x_n) \\ &= [\alpha d(f) + \beta d(g)]\, \Theta(x_1, \cdots x_n) \Rightarrow \\ d(h) &= \alpha d(f) + \beta d(g). \end{aligned}$$

Corollary 1

$$d\begin{pmatrix} \alpha_{11} & \alpha_{12} & \cdots & \alpha_{1n} \\ \vdots & \vdots & & \vdots \\ \alpha\alpha_{k1}+\beta\alpha'_{k1} & \alpha\alpha_{k2}+\beta\alpha'_{k2} & \cdots & \alpha\alpha_{kn}+\beta\alpha'_{kn} \\ \vdots & \vdots & & \vdots \\ \alpha_{n1} & \alpha_{n2} & \cdots & \alpha_{nn} \end{pmatrix}$$

$$= \alpha d\begin{pmatrix} \alpha_{11} & \alpha_{12} & \cdots & \alpha_{1n} \\ \vdots & \vdots & & \vdots \\ \alpha_{k1} & \alpha_{k2} & \cdots & \alpha_{kn} \\ \vdots & \vdots & & \vdots \\ \alpha_{n1} & \alpha_{n2} & \cdots & \alpha_{nn} \end{pmatrix} + \beta d\begin{pmatrix} \alpha_{11} & \alpha_{12} & \cdots & \alpha_{1n} \\ \vdots & \vdots & & \vdots \\ \alpha'_{k1} & \alpha'_{k2} & \cdots & \alpha'_{kn} \\ \vdots & \vdots & & \vdots \\ \alpha_{n1} & \alpha_{n2} & \cdots & \alpha_{nn} \end{pmatrix}$$

Corollary 2

$$d\begin{pmatrix} \alpha_{11} & \alpha_{12} & \cdots & \alpha_{1n} \\ \vdots & \vdots & & \vdots \\ \alpha\alpha_{k1} & \alpha\alpha_{k2} & \cdots & \alpha\alpha_{kn} \\ \vdots & \vdots & & \vdots \\ \alpha_{n1} & \alpha_{n2} & \cdots & \alpha_{nn} \end{pmatrix} = \alpha d\begin{pmatrix} \alpha_{11} & \alpha_{12} & \cdots & \alpha_{1n} \\ \vdots & \vdots & & \vdots \\ \alpha_{k1} & \alpha_{k2} & \cdots & \alpha_{kn} \\ \vdots & \vdots & & \vdots \\ \alpha_{n1} & \alpha_{n2} & \cdots & \alpha_{nn} \end{pmatrix}$$

Corollary 3

$$d\begin{pmatrix} \alpha_{11} & \alpha_{12} & \cdots & \alpha_{1n} \\ \vdots & \vdots & & \vdots \\ \alpha_{i1} & \alpha_{i2} & \cdots & \alpha_{in} \\ \vdots & \vdots & & \vdots \\ \alpha_{j1}+\alpha\alpha_{i1} & \alpha_{j2}+\alpha\alpha_{i2} & \cdots & \alpha_{jn}+\alpha\alpha_{in} \\ \vdots & \vdots & & \vdots \\ \alpha_{n1} & \alpha_{n2} & \cdots & \alpha_{nn} \end{pmatrix} = d\begin{pmatrix} \alpha_{11} & \cdots & \alpha_{1n} \\ \vdots & & \vdots \\ \alpha_{i1} & \cdots & \alpha_{in} \\ \vdots & & \vdots \\ \alpha_{j1} & \cdots & \alpha_{jn} \\ \vdots & & \vdots \\ \alpha_{n1} & \cdots & \alpha_{nn} \end{pmatrix}$$

(since the other summand has, after extracting α, two identical rows).

Let $(\alpha_{ij}) \in D_n$. Put $x_i = (\alpha_{i1}, \cdots, \alpha_{in})$, the ith row of (α_{ij}) viewed as a vector in D^n. We consider d now as a function $d:(x_1, \cdots, x_n) \to D$ of $(D^n)^n$ into D. If $x_i = x_j$ for some $i \neq j$, then $d(x_1, \cdots, x_n) = d(f) = 0$, since the resulting f has a nonzero kernel. Corollary 1 above asserts that d is n-linear. Also, d is nontrivial by the Corollary to Theorem 4. If (α_{ij}) is the identity matrix in D_n, then $d((\alpha_{ij})) = 1$, by definition. Invoking Theorem 2(b), we have

Theorem 5

There is a unique function $d: D_n \to D$ satisfying

(a) d is linear in each row,
(b) if two rows are equal in (α_{ij}), then $d((\alpha_{ij})) = 0$, and
(c) $d(I) = 1$, where I is the identity matrix in D_n.

Corollary

The definition of $d((\alpha_{ij}))$ in terms of a vector space V is independent of the choice of V.

EXAMPLE 1

$$d\begin{pmatrix} 1 & 0 & 1 \\ 0 & 1 & 0 \\ 0 & 0 & 1 \end{pmatrix} = d\begin{pmatrix} 1 & 0 & 0 \\ 0 & 1 & 0 \\ 0 & 0 & 1 \end{pmatrix} + d\begin{pmatrix} 0 & 0 & 1 \\ 0 & 1 & 0 \\ 0 & 0 & 1 \end{pmatrix} = 1 + 0 = 1.$$

EXAMPLE 2

$$d\begin{pmatrix} 2 & 0 & 0 \\ 0 & 3 & 0 \\ 0 & 0 & 4 \end{pmatrix} = 2d\begin{pmatrix} 1 & 0 & 0 \\ 0 & 3 & 0 \\ 0 & 0 & 4 \end{pmatrix} = 2 \cdot 3d\begin{pmatrix} 1 & 0 & 0 \\ 0 & 1 & 0 \\ 0 & 0 & 4 \end{pmatrix}$$

$$= 2 \cdot 3 \cdot 4 \cdot d\begin{pmatrix} 1 & 0 & 0 \\ 0 & 1 & 0 \\ 0 & 0 & 1 \end{pmatrix} = 24.$$

EXERCISES

1. Let $\Theta_1, \cdots, \Theta_m$ be k-linear forms on ${}_DV$, and let $\alpha_1, \cdots, \alpha_m \in D$. Show that $\alpha_1\Theta_1 + \cdots + \alpha_m\Theta_m$ is a k-linear form on V, where

$$(\alpha_1\Theta_1 + \cdots + \alpha_m\Theta_m)(x_1, \cdots, x_k) = \alpha_1\Theta_1(x_1, \cdots, x_k) + \cdots + \alpha_m\Theta_m(x_1, \cdots, x_k), \text{ for } x_1, \cdots, x_k \in V.$$

2. Let $R = \{f \in A(\mathbb{R}^2) : d(\bar{f}) = 1 \text{ where } f = \bar{f} \cdot t_v\}$. Show that R is a normal subgroup of $A(\mathbb{R}^2)$.

3. (a) Show, using Exercise 1, that the set of alternating bilinear forms on $\mathbb{R}^3$ is a vector space.

   (b) Describe the space of alternating bilinear forms on $\mathbb{R}^3$ in terms of their matrix representations with respect to the standard basis.

4. Let u, v be linearly independent vectors in $\mathbb{R}^3$.

   (a) Show that there exists an alternating bilinear form Θ on $\mathbb{R}^3$ such that $\Theta(u, v) = 1$. Is Θ unique?
   (b) Let $u' = \alpha_1 u + \beta_1 v$ and $v' = \alpha_2 u + \beta_2 v$. Compute $\Theta(u', v')$, using the Θ of part (a).
   (c) Suppose that $\{u, v\}$ are orthonormal. Explain what $\Theta(u', v')$ should represent, using the notation of parts (a) and (b).

5. Fill in the details of the discussion of area preceding Proposition 1. In particular:

   (a) show how Exercise 4, Section 3.5 may be utilized;
   (b) indicate how to define Θ, and hence Θ', on all of $\mathbb{R}^3 \times \mathbb{R}^3$;
   (c) show some parallelogram whose area is not given by Θ';
   (d) compute Θ' explicitly when $M = \langle(1, 1, 1), (1, -\frac{1}{2}, -\frac{1}{2})\rangle$.

6. Let $\dim(V) = n$, and let $f \in \operatorname{Hom}_D(V, V)$. Discuss how $d(f)$ can be interpreted as the "stretching factor" of the map f as it acts on n-dimensional parallelepipeds. Include some examples of maps on $\mathbb{R}^2$ and $\mathbb{R}^3$, showing how they affect the area (respectively volume) of the unit square (respectively unit cube).

7. Using determinants, discuss what effect an isomorphism of $\mathbb{R}^2 \to \mathbb{R}^2$ has on an ellipse and on a hyperbola:

   (a) in terms of the "discriminant" of the given conic;
   (b) in terms of the quadratic form interpretation in Section 3.4.

SECTION 3.7
FURTHER PROPERTIES OF DETERMINANTS

Although we have proved the existence and uniqueness of a determinant function $d: D_n \to D$, we have not developed a suitable method of computing the value of d for some specific matrix. In practice one uses some type of reduction process: either reducing the given $n \times n$ matrix to another $n \times n$ matrix of simpler form or reducing the problem to computing the value of d on $m \times m$ matrices where $m < n$. We shall indicate a method for each of these types of reduction.

Throughout this discussion, we shall use the symbol d to denote the determinant function on any size matrix; the context should make it clear which particular function we have in mind.

Recall the definitions of the elementary matrices $E(\lambda; i)$, $E(\lambda; i, j)$, and $E(i, j)$. We observe that

$$E(\lambda; i)^T = E(\lambda; i),$$
$$E(\lambda; i, j)^T = E(\lambda; j, i),$$
$$E(i, j)^T = E(i, j).$$

Lemma 1

Let $n \geq 2$. In D_n we have

(a) $d(E(\lambda; i)) = \lambda$,
(b) $d(E(\lambda; i, j)) = 1$, and
(c) $d(E(i, j)) = -1$. Moreover, $d(E^T) = d(E)$ for any elementary matrix E.

Proof

(a) Using linearity in the ith row, $d(E(\lambda; i)) = \lambda d(I) = \lambda$.
(b) Using linearity in the ith row, $d(E(\lambda; i, j)) = d(I) + d(B)$, where the ith row of B is λ times the jth row of B; hence $d(B) = 0$.
(c) Since $E(i, j)$ is obtained from I by interchanging the ith and jth rows, and since d is alternating, $d(E(i, j)) = -1d(I) = -1$. The final assertion of the lemma follows from (a), (b), and (c) together with the observation preceding the lemma.

Theorem 1

Let $A \in D_n$. Then $d(A) = d(A^T)$.

Proof

If A is singular so is A^T, so $d(A) = d(A^T) = 0$. If not, decompose A into a product $A = E_1 \cdots E_k$ of elementary matrices. Then $A^T = E_k{}^T \cdots E_1{}^T$.

Corollary

The function $d: D_n \to D$, when viewed as a function on the columns of a matrix, is alternating and n-linear.

This corollary asserts that in determinant theory, each theorem concerning the rows of a matrix has a corresponding theorem about the columns.

Lemma 2

If

$$A = \begin{pmatrix} 1 & \alpha_{12} & \cdots & \cdots & \alpha_{1n} \\ 0 & 1 & \alpha_{23} & \cdots & \alpha_{2n} \\ 0 & 0 & & & \\ \vdots & & \ddots & & \vdots \\ 0 & 0 & \cdots & 0 & 1 \end{pmatrix},$$

then $d(A) = 1$.

Proof

By Chapter 1, Section 1.9, Exercise 5, A can be written

$$A = E(\alpha_{n-1,n}; n-1, n) \cdots E(\alpha_{1n}; 1, n)$$

as a product of elementary matrices each of which has determinant 1.

Corollary

If

$$A = \begin{pmatrix} \alpha_{11} & \alpha_{12} & \cdots & \alpha_{1n} \\ 0 & \alpha_{22} & \cdots & \alpha_{2n} \\ \vdots & & & \vdots \\ 0 & \cdots & \cdots & \alpha_{nn} \end{pmatrix},$$

then $d(A) = \alpha_{11}\alpha_{22} \cdots \alpha_{nn}$. In words, the determinant of an upper triangular matrix is the product of its diagonal elements.

Proof

Use linearity in each of the rows to get 1's down the diagonal.

We can now show one reduction technique. Let $A \in D_n$. Write $A = E_1 \cdots E_k A'$, where

$$A' = \begin{pmatrix} \beta_{11} & \beta_{12} & \cdots & \beta_{1n} \\ 0 & \beta_{22} & \cdots & \cdots \\ \vdots & & & \vdots \\ 0 & \cdots & 0 & \beta_{nn} \end{pmatrix},$$

and E_i is elementary. Then $d(A) = d(E_1) \cdots d(E_k) \cdot \beta_{11}\beta_{22} \cdots \beta_{nn}$.

Of course one can always factor A completely into a product of elementary matrices, but the present technique saves half the work. For example, let

$$A = \begin{pmatrix} 2 & 4 & 8 \\ 3 & 7 & 5 \\ 1 & 2 & 1 \end{pmatrix}.$$

Then

$$A = \begin{pmatrix} 2 & 0 & 0 \\ 0 & 1 & 0 \\ 0 & 0 & 1 \end{pmatrix} \cdot \begin{pmatrix} 1 & 2 & 4 \\ 3 & 7 & 5 \\ 1 & 2 & 1 \end{pmatrix} = \begin{pmatrix} 2 & 0 & 0 \\ 0 & 1 & 0 \\ 0 & 0 & 1 \end{pmatrix}\begin{pmatrix} 1 & 0 & 0 \\ 3 & 1 & 0 \\ 0 & 0 & 1 \end{pmatrix}\begin{pmatrix} 1 & 2 & 4 \\ 0 & 1 & -7 \\ 1 & 2 & 1 \end{pmatrix}$$

$$= \begin{pmatrix} 2 & 0 & 0 \\ 0 & 1 & 0 \\ 0 & 0 & 1 \end{pmatrix}\begin{pmatrix} 1 & 0 & 0 \\ 3 & 1 & 0 \\ 0 & 0 & 1 \end{pmatrix}\begin{pmatrix} 1 & 0 & 0 \\ 0 & 1 & 0 \\ 1 & 0 & 1 \end{pmatrix}\begin{pmatrix} 1 & 2 & 4 \\ 0 & 1 & -7 \\ 0 & 0 & -3 \end{pmatrix}.$$

Hence $d(A) = 2 \cdot 1 \cdot 1 \cdot (-3) = -6$.

In practice, the process is streamlined somewhat by thinking of each step as a row (or column) operation, and collecting the values $d(E_i)$ rather than carrying along the E_i's. The above computation would then go as follows:

$$d\begin{pmatrix} 2 & 4 & 8 \\ 3 & 7 & 5 \\ 1 & 2 & 1 \end{pmatrix} = 2d\begin{pmatrix} 1 & 2 & 4 \\ 3 & 7 & 5 \\ 1 & 2 & 1 \end{pmatrix} = 2d\begin{pmatrix} 1 & 2 & 4 \\ 0 & 1 & -7 \\ 1 & 2 & 1 \end{pmatrix} = 2d\begin{pmatrix} 1 & 2 & 4 \\ 0 & 1 & -7 \\ 0 & 0 & -3 \end{pmatrix} = -6.$$

The second technique, called *expansion by cofactors*, arises from an analysis of the induction step by which we proved the existence of alternating k-linear forms.

DEFINITION Let $A = (\alpha_{ij}) \in D_n$. We denote by $A(i \mid j)$ the $(n-1) \times (n-1)$ matrix obtained by deleting the ith row and jth column from A. The scalar $(-1)^{i+j} d(A(i \mid j))$ is called the *cofactor of* (*or corresponding to*) α_{ij}.

Theorem 2

Let $A = (\alpha_{ij}) \in D_n$. Then

$$d(A) = \sum_{i=1}^{n} (-1)^{i+j} \alpha_{ij}\, d(A(i \mid j)).$$

Proof

Let $x_i = (\gamma_{i1}, \cdots, \gamma_{in}) \in D^n$, for $i = 1, \cdots, n-1$. Then $d: D_{n-1} \to D$ determines, for any j, $1 \leq j \leq n$, an alternating $(n-1)$-linear form $\tilde{d}$ on D^n

defined by

$$\tilde{d}(x_1, \cdots, x_{n-1}) = d\begin{pmatrix} \gamma_{11} & \cdots & \gamma_{1\,j-1} & \gamma_{1\,j+1} & \cdots & \gamma_{1n} \\ \vdots & & & & & \\ \gamma_{n-1\,1} & \cdots & \gamma_{n-1\,j-1} & \gamma_{n-1\,j+1} & \cdots & \gamma_{n-1\,n} \end{pmatrix}.$$

If $\{e_1, \cdots, e_n\}$ is the standard basis of D^n, we see that $\tilde{d}$ is nontrivial for $d(e_1, \cdots, \hat{e}_j, \cdots, e_n) = 1$.

Let $f: D^n \to D$ be the unique homomorphism such that

$$(e_i)f = \begin{Bmatrix} 0 & \text{if} & i \neq j \\ 1 & \text{if} & i = j \end{Bmatrix}.$$

Then

$$\Theta(x_1, \cdots, x_n) = \sum_{i=1}^{n} (-1)^{i+1}(x_i)f \cdot \tilde{d}(x_1, \cdots, \hat{x}_i, \cdots, x_n)$$

is an alternating n-linear form on D^n by the proof of Theorem 1, and hence there exists $\lambda \in D$ such that $d = \lambda \cdot \Theta$.

In particular

$$\begin{aligned} 1 = d(I) &= \lambda \cdot \Theta(I) \\ &= \lambda \sum_{i=1}^{n} (-1)^{i+1}(e_i)f \cdot \tilde{d}(e_1, \cdots, \hat{e}_i, \cdots, e_n) \\ &= \lambda \cdot (-1)^{j+1} \cdot 1\tilde{d}(e_1, \cdots, \hat{e}_j, \cdots, e_n) = \lambda(-1)^{j+1} \end{aligned}$$

so

$$\lambda = (-1)^{j+1} = (-1)^{j-1}.$$

Application of this, where $x_i = (\alpha_{i1}, \cdots, \alpha_{in})$ yields

$$\begin{aligned} d((\alpha_{ij})) &= \sum_{i=1}^{n} (-1)^{i+1}(-1)^{j-1}\alpha_{ij}\, d(A(i \mid j)) \\ &= \sum_{i=1}^{n} (-1)^{i+j}\alpha_{ij}\, d(A(i \mid j)). \end{aligned}$$

EXAMPLE 1

$$d\begin{pmatrix} 2 & 3 & 1 \\ 1 & -1 & 2 \\ 6 & 4 & 3 \end{pmatrix} = 2 \cdot d\begin{pmatrix} -1 & 2 \\ 4 & 3 \end{pmatrix} - 1 \cdot d\begin{pmatrix} 3 & 1 \\ 4 & 3 \end{pmatrix} + 6d\begin{pmatrix} 3 & 1 \\ -1 & 2 \end{pmatrix}$$

$$\begin{aligned} &= 2(-3-8) - 1(9-4) + 6(6+1) \\ &= -22 - 5 + 42 = 15. \end{aligned}$$

EXAMPLE 2 Let

$$A = \begin{pmatrix} 2 & 0 & 3 \\ 1 & 2 & 4 \\ 1 & 0 & 2 \end{pmatrix}.$$

Since there are two zeros in the 2d column, we expand along that column:

$$\begin{aligned} d(A) &= -0 \cdot d\begin{pmatrix} 1 & 4 \\ 1 & 2 \end{pmatrix} + 2 \cdot d\begin{pmatrix} 2 & 3 \\ 1 & 2 \end{pmatrix} - 0 \cdot d\begin{pmatrix} 2 & 3 \\ 1 & 4 \end{pmatrix} \\ &= 0 + 2(4-3) - 0 = 2. \end{aligned}$$

EXERCISES

1. Compute the determinant of

$$\begin{pmatrix} 1 & -1 & 1 \\ 2 & 1 & 3 \\ 0 & 1 & 0 \end{pmatrix}, \begin{pmatrix} 2 & 0 & 0 \\ 1 & 2 & 3 \\ -1 & 4 & 1 \end{pmatrix}, \begin{pmatrix} 1 & 2 & 0 & 0 \\ 3 & 4 & 0 & 0 \\ 0 & 0 & 1 & -1 \\ 0 & 0 & 7 & 3 \end{pmatrix},$$

$$\begin{pmatrix} 1 & 2 & 1 & 6 \\ 3 & 4 & 7 & 5 \\ 0 & 0 & 1 & -1 \\ 0 & 0 & 7 & 3 \end{pmatrix}, \begin{pmatrix} 1 & 2 & 0 & 0 \\ 3 & 4 & 0 & 0 \\ 2 & -1 & 1 & -1 \\ 5 & 7 & 7 & 3 \end{pmatrix}.$$

2. Generalize the last three examples of Exercise 1.

3. State a theorem about cofactor expansion along a row, and prove it.

4. Compute

$$d\begin{pmatrix} 2 & 0 & 0 \\ 1 & 1 & 3 \\ -1 & -2 & 4 \end{pmatrix}, d\begin{pmatrix} 1 & 0 & 0 & 0 & 0 \\ 2 & 5 & 7 & 1 & 6 \\ 3 & 0 & 2 & 0 & 0 \\ 4 & 0 & 1 & -1 & 9 \\ 5 & 0 & 3 & 0 & \frac{1}{2} \end{pmatrix}.$$

*5. Let D be a field and $D' \subseteq D$ such that if $\alpha, \beta \in D'$, then $\alpha \pm \beta \in D'$ and $\alpha\beta \in D'$. Suppose $A = (\alpha_{ij}) \in D_n$, with $\alpha_{ij} \in D'$, $\forall i, j$. Prove that $d(A) \in D'$. [*Hint*: Use induction on the size, n, of the matrix and the method of expansion by cofactors.] [3.8]

6. Let V be an inner product space with orthonormal basis B. Let $f: V \to V$ be an isometry, and put $A = m(f, B)$. Find $d(A)$. [*Hint*: By previous results, show $AA^T = I$.]

7. Using Exercise 6, let $B = \{(1, 0), (0, 1)\}$ in $\mathbb{R}^2$, and show that if f is an isometry (with respect to the standard inner product) of $\mathbb{R}^2$, then

$$m(f, B) = \begin{pmatrix} \cos\theta & \sin\theta \\ -\sin\theta & \cos\theta \end{pmatrix}$$

or

$$m(f, B) = \begin{pmatrix} \cos\theta & \sin\theta \\ \sin\theta & -\cos\theta \end{pmatrix},$$

for some θ. In the first case, f is a *rotation*; in the second case f is a *reflection*.

8. Let $A \in D_n$. Show that

$$\sum_{i=1}^{n} (-1)^{i+j} \alpha_{ik}\, d(A(i \mid j)) = 0 \text{ if } k \neq j.$$

[*Hint*: Show that this sum is the determinant of a matrix with two identical columns.]

9. Let $A \in D_n$. Let $d(A) = \delta$. The matrix whose (i, j)th entry is $(-1)^{i+j}$ $d(A(j \mid i))$ is called the *adjoint of A*, denoted adj (A). Prove that

$$A \cdot \operatorname{adj}(A) = \begin{pmatrix} \delta & 0 & \cdots & 0 \\ 0 & \delta & \cdots & 0 \\ \vdots & \vdots & & \vdots \\ 0 & 0 & \cdots & \delta \end{pmatrix} = \delta \cdot I$$

and hence,

$$A^{-1} = \operatorname{adj}(A) \cdot (\delta^{-1} I),$$

provided $\delta \neq 0$. We shall occasionally refer to this adjoint of A as the classical adjoint of A, to distinguish it from another common use of the word adjoint that will be encountered in the next section.

10. (Cramer's rule.) Show that the system of equations

$$\begin{matrix} \alpha_{11}x_1 + \cdots + \alpha_{1n}x_n = \beta_1, \\ \vdots \\ \alpha_{n1}x_1 + \cdots + \alpha_{nn}x_n = \beta_n \end{matrix}$$

has the (unique) solution

$$x_i = \delta^{-1} \sum_{j=1}^{n} (-1)^{i+j} d(A(j\,|\,i)) \cdot \beta_j,\ i = 1, 2, \cdots, n,$$

provided $d(A) = \delta \neq 0$, where $A = (\alpha_{ij})$.

11. Solve the system

$$\begin{aligned} x_1 - x_2 + x_3 &= 2, \\ 2x_1 + x_2 + 3x_3 &= -1, \\ x_1 + x_2 + x_3 &= 1. \end{aligned}$$

SECTION 3.8 APPLICATIONS OF THE DETERMINANT FUNCTION

If D is a field and ${}_DV$ a vector space, the easiest homomorphisms to study are the multiplication maps, that is, for $\alpha \in D$ the map $f_\alpha : V \to V$ defined by $(x)f_\alpha = \alpha x$ for all $x \in V$. In this section we shall investigate the relation between an arbitrary homomorphism $f : V \to V$ and these multiplication maps.

DEFINITION 1 Let $f : {}_DV \to {}_DV$ be a homomorphism and $\lambda \in D$. Set $V_\lambda = \{x \in V : (x)f = \lambda x\}$. If $V_\lambda \neq O$, λ is called an *eigenvalue* of f and a nonzero element of V_λ is called an *eigenvector* of f (relative to λ), and V_λ is called the *eigenspace* of λ.

Proposition 1

In the notation of Definition 1, $V_\lambda < V$.

Proof

If $x, y \in V_\lambda$, $\alpha, \beta \in D$, then

$$\begin{aligned} (\alpha x + \beta y)f &= \alpha(x)f + \beta(y)f = \alpha \cdot \lambda \cdot x + \beta \cdot \lambda \cdot y \\ &= \lambda\alpha x + \lambda\beta y = \lambda(\alpha x + \beta y). \end{aligned}$$

Therefore $\alpha x + \beta y \in V_\lambda$ and $V_\lambda < V$.

It might be noted that $V_0 = \text{Ker}\,(f)$, but we need a technique for finding nonzero $\lambda \in D$ such that $V_\lambda \neq O$. Happily, determinants provide us with such a technique. Clearly

$$V_\lambda \neq O \Leftrightarrow \text{Ker}\,(f - f_\lambda) \neq O \Leftrightarrow d(f - f_\lambda) = 0.$$

We need to extend our definition of determinants to include the determinant of a matrix whose entries are polynomials $p(X)$ in X with coefficients in D. To do so, let

$$F = \{p(X)/q(X) : p(X), q(X) \in D[X], q(X) \neq 0, p(X), q(X) \text{ have no common factors}\}.$$

Then F is a field under the usual operations of addition and multiplication of polynomials. Moreover, F contains the subfield of constant polynomials that we identify with D, as well as the subset where $q(X) = 1$ that we identify with $D[X]$. If $A \in F_n$, then $d(A)$ makes sense, and belongs to F. By Exercise 5, Section 3.7, if all the entries of A belong to D (respectively $D[X]$), then $d(A) \in D$ (respectively $D[X]$).

EXAMPLE 1

$$d\begin{pmatrix} 1-X & 2 \\ 3 & 2-X \end{pmatrix} = (1-X)(2-X) - 6 = X^2 - 3X - 4 \in \mathbb{R}[x].$$

EXAMPLE 2

$$\begin{aligned} d\begin{pmatrix} -X & 0 & 1 \\ 0 & 2-X & 0 \\ 1 & 0 & 2+X \end{pmatrix} &= (-X)d\begin{pmatrix} 2-X & 0 \\ 0 & 2+X \end{pmatrix} + 1d\begin{pmatrix} 0 & 1 \\ 2-X & 0 \end{pmatrix} \\ &= (-X)(2-X)(2+X) + 1(X-2) \\ &= X^3 - 3X - 2 \in \mathbb{R}[X]. \end{aligned}$$

DEFINITION 2 Let ${}_DV$ be a vector space, $B = \{x_1, \cdots, x_n\}$ a basis of V, and $f: V \to V$ a homomorphism. Let $(\alpha_{ij}) = m(f, B)$. The value $d((\alpha_{ij}) - XI_n)$ is a polynomial of degree less than or equal to n in X called the *characteristic polynomial of f*. We also call this the *characteristic polynomial of the matrix* $A = (\alpha_{ij})$. Exercise 1 shows that this polynomial does not depend on the choice of B.

Our earlier remarks now constitute a proof of:

Proposition 2

Let $f: {}_DV \to {}_DV$ be a homomorphism on the finite dimensional vector space V. Then $\lambda \in D$ is an eigenvalue of $f \Leftrightarrow \lambda$ is a root of the characteristic polynomial of f.

Proposition 3

Let ${}_DV$ be finite dimensional, $f: V \to V$ a homomorphism, and $\lambda_1, \cdots, \lambda_p$ distinct eigenvalues of f. Then

$$V_{\lambda_1} + \cdots + V_{\lambda_p} = V_{\lambda_1} \oplus \cdots \oplus V_{\lambda_p}.$$

Proof

The proof is by induction on p. If $p = 2$ let $x \in V_{\lambda_1} \cap V_{\lambda 2}$. Then

$$(x)f = \lambda_1 x = \lambda_2 x \Rightarrow O = (\lambda_1 - \lambda_2)x \text{ and } \lambda_1 \neq \lambda_2 \Rightarrow x = O.$$

Now assume the result is true for $p - 1$ distinct eigenvalues ($p \geq 3$). If for i, $1 \leq i \leq p$,

$$x_i = \sum_{\substack{j=1 \\ j \neq i}}^{p} x_j, \; (x_i)f = \sum_{\substack{j=1 \\ j \neq i}}^{p} (x_j)f \Rightarrow \sum_{\substack{j=1 \\ j \neq i}}^{p} (\lambda_j - \lambda_i)x_j = O \Rightarrow \text{(since } \lambda_j - \lambda_i \neq 0)$$

$$x_k = \sum_{\substack{j=1 \\ j \neq i, k}}^{p} (\lambda_k - \lambda_i)^{-1}(\lambda_j - \lambda_i)x_j,$$

contradicting the induction hypothesis, and so

$$V_{\lambda_1} + \cdots + V_{\lambda_p} = V_{\lambda_1} \oplus \cdots \oplus V_{\lambda_p}.$$

Corollary

If $\dim {}_D V = n$, and $f: V \to V$, then f has at most n eigenvalues.

Proposition 4

Let ${}_D V$ be a finite dimensional vector space and $f: V \to V$ a homomorphism whose distinct eigenvalues are $\lambda_1, \cdots, \lambda_p$. Then there exists a basis B of V such that $m(f, B)$ is diagonal $\Leftrightarrow \sum_{i=1}^{p} \dim (V_{\lambda_i}) = \dim V$.

Proof

The proof is left as an exercise for the reader.

To this point we have been somewhat cavalier about eigenvalues in that we have not discussed their existence. If ${}_{\mathbb{R}}V$ is a real vector space of dimension n and $f: V \to V$ is a homomorphism, the characteristic polynomial of f is a polynomial of degree n with coefficients in $\mathbb{R}$. Such polynomials have roots in $\mathbb{C}$ but not necessarily in $\mathbb{R}$. Moreover, even if the characteristic polynomial of f has n roots in $\mathbb{R}$, the roots might not be distinct so we would have no guarantee that V has a basis of eigenvectors. We shall now establish necessary and sufficient conditions for an endomorphism to have a basis of eigenvectors; however, we need first to define a new map f^* induced by a map f on an inner product space.

Suppose V is such a space with an orthonormal basis $\{x_1, \cdots, x_n\}$. Let $f \in \operatorname{Hom}_{\mathbb{R}}(V, V)$ and y be a fixed vector in V. We define $f_y: V \to \mathbb{R}$ by $(x)f_y = (x)f \cdot y$.

Since

$$\begin{aligned}(\alpha x + \beta x')f_y &= (\alpha x + \beta x')f \cdot y \\ &= (\alpha(x)f + \beta(x')f) \cdot y \\ &= \alpha(x)f \cdot y + \beta(x')f \cdot y \\ &= \alpha(x)f_y + \beta(x')f_y,\end{aligned}$$

we have $f \in \operatorname{Hom}_{\mathbb{R}}(V, \mathbb{R})$.

Setting $(x_i)f \cdot y = \alpha_i$, $i = 1, 2, \cdots, n$, and $y^* = \sum_{i=1}^{n} \alpha_i x_i$, we observe that the function $x \to x \cdot y^*$ is also linear, and, moreover, it agrees with f_y on the basis $\{x_1, \cdots, x_n\}$. Thus with each $y \in V$ we have associated a vector $y^* \in V$ such that the map $x \to (x)f \cdot y$ is the same as the map $x \to x \cdot y^*$. We give the function $y \to y^*$ the name f^*, that is, $(y)f^* = y^*$, and claim that $f^* \in \operatorname{Hom}_{\mathbb{R}}(V, V)$. For suppose $y, y' \in V$ and $\alpha, \beta \in \mathbb{R}$. Then for each $x \in V$,

$$\begin{aligned}x \cdot (\alpha y + \beta y')f^* &= (x)f \cdot (\alpha y + \beta y') \\ &= \alpha(x)f \cdot y + \beta(x)f \cdot y' \\ &= \alpha(x \cdot (y)f^*) + \beta(x \cdot (y')f^*) \\ &= x \cdot \alpha(y)f^* + x \cdot \beta(y')f^* \\ &= x \cdot (\alpha(y)f^* + \beta(y')f^*).\end{aligned}$$

Since x was arbitrary,

$$(\alpha y + \beta y')f^* = \alpha(y)f^* + \beta(y')f^*,$$

yielding $f^* \in \operatorname{Hom}_{\mathbb{R}}(V, V)$. The map f^* is called the *adjoint* of f (not to be confused with the classical adjoint defined in the previous section). It is characterized by

$$(x)f \cdot y = x \cdot (y)f^*, \ \forall x, y \in V.$$

Of course we can perform the same construction for $f \in \operatorname{Hom}_{\mathbb{C}}(V, V)$ if ${}_{\mathbb{C}}V$ is a vector space with a unitary form (see Exercise 5, Section 3.5), that is, for a basis $\{x_1, \cdots, x_n\}$ we have

$$\left(\sum_{i=1}^{n} \alpha_i x_i\right) \cdot \left(\sum_{i=1}^{n} \beta_i x_i\right) = \sum_{i=1}^{n} \alpha_i \beta_i^{\sigma},$$

where σ denotes (complex) conjugation.

Proposition 5

(a) Let ${}_{\mathbb{R}}V$ have orthonormal basis B and let $f \in \operatorname{Hom}(V, V)$. Then

$$m(f^*, B) = m(f, B^T).$$

(b) Let ${}_{\mathbb{C}}V$ have orthornormal basis B and let $f \in \operatorname{Hom}(V, V)$. Then

$$m(f^*, B) = m(f, B)^{T\sigma}.$$

Proof

We prove (a) and leave part (b) to the reader.

To say $m(f, B) = (\alpha_{ij})$ means that

$$(x_i)f = \sum_{j=1}^{n} \alpha_{ij} x_j, \; i = 1, \cdots, n.$$

Then $x_i \cdot (x_j)f^* = (x_i)f \cdot x_j = \alpha_{ij}$ for each i and j. But if

$$(x_j)f^* = \sum_{k=1}^{n} \beta_{jk} x_k,$$

(that is, $m(f^*, B) = (\beta_{ij})$), then

$$\beta_{ji} = x_i \cdot \sum_{k=1}^{n} \beta_{jk} x_k = x_i \cdot (x_j)f^* = \alpha_{ij},$$

and the conclusion follows.

Referring to Chapter 1, Section 1.8, Exercise 8, we can compare the eigenvalues of a map $f \in \operatorname{Hom}_{\mathbb{R}}(V, V)$ with the eigenvalues of the map $\hat{f} \in \operatorname{Hom}_{\mathbb{C}}(\hat{V}, \hat{V})$ on the complexification $\hat{V}$ of V. To begin with, if $(x)f = \lambda x$ then

$$(x + 0i)\hat{f} = (x)f + (0)fi = \lambda x + 0i = \lambda(x + 0i).$$

Thus each eigenvalue of f is also an eigenvalue of $\hat{f}$. The next result shows that whenever λ is an eigenvalue of $\hat{f}$ and λ is real, then λ is also an eigenvalue of f.

Lemma

Let f and $\hat{f}$ be as above. If λ is an eigenvalue of $\hat{f}$ and λ is real, then λ is also an eigenvalue of f.

Proof

Suppose $(x + yi)\hat{f} = \lambda(x + yi)$. Then

$$\lambda x + \lambda yi = (x + yi)\hat{f} = (x)f + (y)fi,$$

whence $\lambda x = (x)f$ and $\lambda y = (y)f$, so that λ is an eigenvalue of f with corresponding eigenvector x (or y in case $x = O$; not both x and y can be O).

Proposition 6

Let ${}_{\mathbb{C}}V$ have a unitary form denoted by "$\cdot$" and $g \in \operatorname{Hom}_{\mathbb{C}}(V, V)$. If $g = g^*$, then all the eigenvalues of g are real.

Proof

Suppose $(x)g = \lambda x$ where $x \neq O$. Without loss of generality we assume $x \cdot x = 1$. Then

$$\lambda = \lambda x \cdot x = (x)g \cdot x = x \cdot (x)g^* = x \cdot (\lambda x) = \lambda^\sigma(x \cdot x) = \lambda^\sigma;$$

thus $\lambda = \lambda^\sigma$ is real. (σ denotes complex conjugation.)

The Fundamental Theorem of Algebra tells us that a polynomial equation with complex coefficients has a root in $\mathbb{C}$ so each $f \in \text{Hom}_{\mathbb{C}}(V, V)$ has at least one eigenvalue in $\mathbb{C}$. This allows us to prove the existence of eigenvalues for certain maps in $\text{Hom}_{\mathbb{R}}(V, V)$.

Proposition 7

Let $f \in \text{Hom}_{\mathbb{R}}(V, V)$ with $f = f^*$. Then f has at least one (real) eigenvalue.

Proof

Let $\hat{f} \in \text{Hom}_{\mathbb{C}}(\hat{V}, \hat{V})$ be the induced map on the complexification ${}_{\mathbb{C}}\hat{V}$ of ${}_{\mathbb{R}}V$. Let B and $\hat{B}$ be corresponding bases for ${}_{\mathbb{R}}V$ and ${}_{\mathbb{C}}\hat{V}$. By Chapter 1, Section 1.1, Exercise 8,

$$A = m(f, B) = m(\hat{f}, \hat{B}).$$

By Section 3.5, Exercise 6, if B is chosen to be orthonormal then $\hat{B}$ is also orthonormal. Since $A^T = A^{T\sigma}$ (the entries are all real), we have

$$m(\hat{f}^*, \hat{B}) = A^{T\sigma} = A^T = m(\hat{f}, \hat{B}),$$

so $\hat{f}^* = \hat{f}$. By Proposition 6 the eigenvalues of $\hat{f}$ are real. Moreover the eigenvalues of $\hat{f}$ are exactly the roots of the equation $d(A - XI) = 0$, and so, by the Fundamental Theorem of Algebra, at least one root exists. Finally, by the lemma the eigenvalues of $\hat{f}$ are also eigenvalues of f.

We can now culminate this development with the promised characterization of those endomorphisms that yield a basis of eigenvectors.

Theorem 1 (Real Spectral Theorem)

Let ${}_{\mathbb{R}}V$ be an inner product space, and let $f \in \text{Hom}_{\mathbb{R}}(V, V)$. There exists a basis of V consisting of eigenvectors for f if and only if $f = f^*$.

Proof

The proof is by induction on the dimension of V. The result obviously holds if the dimension is 1, so we assume that it holds for all vector spaces of dimension less than n and that $\dim(V) = n$.

By Proposition 6 there exists $\lambda \in \mathbb{R}$ and $x \in V$, $x \neq O$ such that $(x)f = \lambda x$. Thus $V_\lambda \neq O$ and $V = V_\lambda \oplus V_\lambda^\perp$ with $\dim (V_\lambda^\perp) < \dim (V)$. We claim that f maps $V_\lambda^\perp$ into itself. Suppose $y \in V_\lambda^\perp$. Then for $x \in V_\lambda$ we have

$$x \cdot (y)f = (x)f \cdot (y) = (\lambda)x \cdot y = \lambda(x \cdot y) = 0.$$

Since x was chosen arbitrarily, we have $(y)f \in V_\lambda^\perp$ as predicted.

The restriction of f to $V_\lambda^\perp$ is clearly self-adjoint, that is, is equal to its adjoint. Therefore, by induction hypothesis $V_\lambda^\perp$ has a basis of eigenvectors for this restriction, and the elements of this basis are also eigenvectors for f. Extending this basis to a basis for V by adding a basis of V_λ gives the desired result.

We conclude this section with a remarkable theorem due to Cayley and Hamilton. We first must establish some facts about $\operatorname{Hom}_D(V, V)$ and D_n.

Earlier we observed that $\operatorname{Hom}_D(V, V)$ is an Abelian group under addition. At the time, D was not necessarily *commutative*. Since D is now commutative we observe that $\operatorname{Hom}_D(V, V)$ can be made into a vector space over D by defining, for $\alpha \in D$, $f \in \operatorname{Hom}_D(V, V)$, the map $\alpha f: V \to V$ by

$$(x)(\alpha f) = \alpha \cdot (x)f, \ \forall x \in V.$$

Letting $B = \{x_1, \cdots, x_n\}$ be a basis of ${}_DV$, we define $f_{ij}: V \to V, 1 \leq i, j \leq n$ by

$$(x_k)f_{ij} = \begin{Bmatrix} x_j & \text{if } k = i \\ 0 & \text{if } k \neq i \end{Bmatrix}.$$

It is a simple exercise to show that $\{f_{ij}: 1 \leq i, j \leq n\}$ is a basis of $\operatorname{Hom}_D(V, V)$, so that the dimension of $\operatorname{Hom}_D(V, V)$ is n^2. Using the correspondence $f \to m(f, B)$, we assert that D_n is likewise a vector space over D where scalar multiplication is given by $\alpha(\alpha_{ij}) = (\alpha\alpha_{ij})$. The "matrix unit" e_{ij} (with 1 in the (i-j)th position and 0 elsewhere) corresponds to f_{ij} so that $\{e_{ij} : 1 \leq i, j \leq n\}$ is a basis of ${}_DD_n$.

Suppose that $f \in \operatorname{Hom}_D(V, V)$. By f^k, we mean $f \cdot f \cdot \cdots \cdot f$ (k-factors). Consider the $n^2 + 1$ elements $1, f, f^2, \cdots, f^{n^2}$ in $\operatorname{Hom}_D(V, V)$. Since $\operatorname{Hom}_D(V, V)$ has dimension n, $\exists \beta_0, \beta_1, \cdots, \beta_{n^2} \in D$ such that

$$\beta_0 \cdot 1 + \beta_1 f + \cdots + \beta_{n^2} f^{n^2} = 0.$$

In other words, *each f satisfies a polynomial of degree less than or equal to n^2 over D*. (Likewise if $A \in D_n$, then $\exists \beta_0, \beta_1, \cdots, \beta_{n^2} \in D$ such that

$$\beta_0 I + \beta_1 A + \cdots + \beta_{n^2} A^{n^2} = 0$$

(the zero matrix).)

EXAMPLE

$$\begin{pmatrix} 1 & 1 \\ 0 & 1 \end{pmatrix}^2 = \begin{pmatrix} 1 & 2 \\ 0 & 1 \end{pmatrix}$$

so that

$$1 \cdot \begin{pmatrix} 1 & 0 \\ 0 & 1 \end{pmatrix} - 2 \cdot \begin{pmatrix} 1 & 1 \\ 0 & 1 \end{pmatrix} + \begin{pmatrix} 1 & 1 \\ 0 & 1 \end{pmatrix}^2 = \begin{pmatrix} 0 & 0 \\ 0 & 0 \end{pmatrix}.$$

The surprising fact is that, as in the example, each $A \in D_n$ satisfies a polynomial of degree less than or equal to n.

Theorem 2 (Cayley-Hamilton Theorem)

Let $f \in \text{Hom}_D(V, V)$. Let $q(X)$ be the characteristic polynomial of f. Then $q(f) = 0$, when interpreted as above. In terms of matrices, let $A \in D_n$, and let $q(X)$ be the characteristic polynomial of A; then $q(A) = 0$.

Before we prove the theorem, we need an observation about matrices with polynomial entries. Suppose $C = (\gamma_{ij})$ is a matrix with entries from $D[X]$. Suppose that $t \geq 0$ is the highest power of X appearing in any entry of C. Let

$$\gamma_{ij} = \gamma_{ij}^{(t)}X^t + \gamma_{ij}^{(t-1)}X^{t-1} + \cdots + \gamma_{ij}^{(1)}X + \gamma_{ij}^{(0)}.$$

Put $C^{(k)} = (\gamma_{ij}^{(k)})$, for $k = 0, 1, \cdots, t$. Then

$$C = C^{(t)}(X^tI) + C^{(t-1)}(X^{t-1}I) + \cdots + C^{(1)}(XI) + C^{(0)}.$$

EXAMPLE

$$\begin{pmatrix} X^2+1 & 2X^2-3X \\ -2X & X-5 \end{pmatrix} = \begin{pmatrix} X^2 & 2X^2 \\ 0 & 0 \end{pmatrix} + \begin{pmatrix} 0 & -3X \\ -2X & X \end{pmatrix} + \begin{pmatrix} 1 & 0 \\ 0 & -5 \end{pmatrix}$$

$$= \begin{pmatrix} 1 & 2 \\ 0 & 0 \end{pmatrix}\begin{pmatrix} X^2 & 0 \\ 0 & X^2 \end{pmatrix} + \begin{pmatrix} 0 & -3 \\ -2 & 1 \end{pmatrix}\begin{pmatrix} X & 0 \\ 0 & X \end{pmatrix} + \begin{pmatrix} 1 & 0 \\ 0 & -5 \end{pmatrix}.$$

Proof of Theorem 2

We do the matrix version. Recall the formula $C \cdot \text{adj}(C) = d(C) \cdot I$, which holds for any $C \in D_n$ (where adj (C) is the classical adjoint of C). Suppose $q(X) = X^n + \beta_1 X^{n-1} + \cdots + \beta_{n-1}X + \beta_n$ is the characteristic polynomial of A; then

$$(A - XI) \cdot \text{adj}(A - XI) = d(A - XI) \cdot I = q(X) \cdot I.$$

Now the entries of adj $(A - XI)$ are plus or minus determinants of $(n-1) \times (n-1)$ submatrices of $(A - XI)$, and hence polynomials in X of degree at most $n - 1$ (see Exercise 2). Let us write, therefore,

$$\text{adj}(A - XI) = C^{(n-1)}(X^{n-1}I) + C^{(n-2)}(X^{n-2}I) + \cdots + C^{(1)}(XI) + C^{(0)}.$$

Rewriting the first equation, we have:

$$(A - XI)[(C^{(n-1)})(X^{n-1}I) + C^{(n-2)}(X^{n-2}I) + \cdots + C^{(1)}(XI) + C^{(0)}]$$
$$= (X^n + \beta_1 X^{n-1} + \cdots + \beta_n)I.$$

Expanding, and equating coefficients, we have

$$\begin{aligned} C^{(n-1)} &= I, \\ AC^{(n-1)} - C^{(n-2)} &= \beta_1 I, \\ AC^{(n-2)} - C^{(n-3)} &= \beta_2 I, \\ &\vdots \\ AC^{(1)} - C^{(0)} &= \beta_{n-1} I, \\ AC^{(0)} &= \beta_n I. \end{aligned}$$

Now multiply these equations by A^n, A^{n-1}, $\cdots$, A, I respectively, and add to get

$$0 = A^n + \beta_1 A^{n-1} + \cdots + \beta_{n-1} A + \beta_n I = q(A).$$

EXERCISES

1. Let $f \in \text{Hom}_D(V, V)$ with $\dim V = n$. Suppose B, B' are bases of V, and that $m(f, B) = A$, $m(f, B') = A'$. Show that $d(A - XI) = d(A' - XI)$. [*Hint*: Show $\exists P \in D_n$ with $P^{-1}AP = A'$.]

2. Suppose A is an $n \times n$ matrix each of whose entries is a polynomial in X of degree less than or equal to 1. Show that $d(A)$ is a polynomial in X of degree less than or equal to n.

3. Let $A = (\alpha_{ij}) \in D_n$. Define the *trace* of A, denoted tr (A), by tr $(A) = \alpha_{11} + \alpha_{22} + \cdots + \alpha_{nn}$.
   (a) Show that tr $(AC) =$ tr (CA), A, $C \in D_n$.
   (b) Show that tr $(PAP^{-1}) =$ tr (A), for each invertible $P \in D_n$.
   (c) Show that if $A \in D_n$, then the characteristic polynomial of f where $m(f, B) = A$ for some basis B of ${}_DV$ is
   $$(-1)^n X^n + (-1)^{n-1} \operatorname{tr}(A) X^{n-1} + \cdots + d(A).$$

4. Find the characteristic polynomials and the eigenvalues of

   (a) $\begin{pmatrix} 2 & 1 & 0 \\ 0 & 1 & 2 \\ 0 & 2 & 1 \end{pmatrix}$; (b) $\begin{pmatrix} 1 & 0 & 1 \\ 2 & 0 & 2 \\ 1 & -1 & 2 \end{pmatrix}$; (c) $\begin{pmatrix} \alpha & \beta \\ 0 & 0 \end{pmatrix}$.

   View each matrix as having entries in $\mathbb{R}$.

5. Show that the map $f_{ij} \to e_{ij}$ is a vector space isomorphism of $\text{Hom}_D(V, V)$ into D_n, where $\dim {}_DV = n$.

6. Generalize Exercise 4(c).

7. Find the eigenvectors corresponding to 4(a). [*Hint*: Assume the given matrix represents f relative to the standard basis of $\mathbb{R}^3$.]

8. Let λ be an eigenvalue of A (that is, of some f represented by A). Let $g(X)$ be any polynomial with coefficients in D. Show that $g(\lambda)$ is an eigenvalue of $g(A)$. [*Hint*: Show that $A - \lambda I$ divides $g(A) - g(\lambda)I$.]

9. The Spectral Theorem for $\mathbb{C}$ asserts:
   Let ${}_{\mathbb{C}}V$ be an n-dimensional Hermitian space with σ being conjugation. Let $f: V \to V$, and let $m(f, B) = A \in \mathbb{C}_n$ for some basis B of V. Then $A(A^T)^\sigma = (A^T)^\sigma A \Leftrightarrow V$ has a basis of eigenvectors of f.

   Although we do not invite you to prove this theorem, you can show:

   (a) if $(A^T)^\sigma = A$, then the above conclusion holds;
   (b) the above condition is not appropriate for $\mathbb{R}$: $\exists A \in \mathbb{R}_2$ with $A \cdot A^T = A^T A$ but $d(A - XI)$ has no solutions in $\mathbb{R}$.

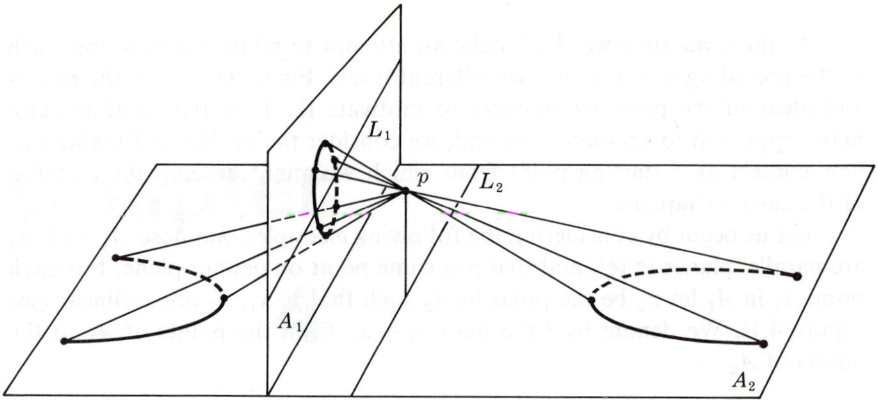

Figure 4.2

parallelism is responsible for the distinction between Figure 4.1 and Figure 4.2. The reader might recall that the occurrence of parallelism has, in the past, caused us to consider special cases—notably in the theorems of Desargues and Ceva.

Projective geometry eliminates the possibility of parallel lines in a plane. As we mentioned in Chapter 2, Sections 2.4 and 2.5, we can enlarge an affine plane by adjoining "points at infinity"—one point for each class of parallel lines. This process, when done carefully, results in a new geometric structure which is an example of a projective plane.

In terms of Figure 4.2, we would then let f map each point at infinity in A_1 to the appropriate point of L_2. Likewise, each point in L_1 would be sent to the appropriate point at infinity on A_2. The extended map would then be 1-1, onto, and acknowledge the geometric structure. In Figure 4.1, the result of adding points at infinity would be that A_1 and A_2 would intersect in precisely their line at infinity. We would merely decree that f leave this line pointwise fixed.

One important feature of enlarging an affine structure to a projective structure is that, once it is done, the points at infinity should be treated exactly as the other points. The situation is analogous to enlarging the field of rational numbers to the field of real numbers via Dedekind cuts; once the new field is constructed, the real numbers that happen not to be rational are treated just like the rationals in regard to the field axioms they satisfy.

One final comment seems desirable in regard to the content of this chapter. Since the preceding material has studied affine geometry but ignored projective geometry, the reader might well ask why we axiomatize projective, rather than affine, geometry. To some extent this is a matter of taste rather than necessity, but we can give two reasons for our choice. First, the projective axioms appear to us the simpler set. Second, although one can go from an affine space to a projective space or from a projective space to an affine space, the second process seems to be the easier from a conceptual point of view in the framework of axiomatic geometry.

3. Complete the proof of Proposition 2 by dealing with the special cases.

4. Let $A \in \mathcal{F}(\mathcal{P})$, p-dim $(A) = 2$. If $L_1, L_2 \in \Lambda$, $L_1 \subseteq A$, $L_2 \subseteq A$, show that $L_1 \cap L_2 \in \mathcal{P}$, that is, coplanar lines intersect. Show also that any two lines which intersect are coplanar.

5. Prove Theorem 1.

6. Prove Proposition 6 in the case $A \cap B = \varnothing$.

*7. Prove Desargues' theorem in the following special case. Let $P, P_1, P_2, P_3, Q_1, Q_2, Q_3$ be distinct points, no four of which lie on a line, let P_1, P_2, P_3 be noncollinear and Q_1, Q_2, Q_3 be noncollinear, and let P, P_1, P_2, P_3 be noncoplanar. Prove that if

$$P = (P_1 + Q_1) \cap (P_2 + Q_2) \cap (P_3 + Q_3),$$

then

$$R_1 = (P_2 + P_3) \cap (Q_2 + Q_3),\ R_2 = (P_1 + P_3) \cap (Q_1 + Q_3),$$
$$R_3 = (P_1 + P_2) \cap (Q_1 + Q_2)$$

are collinear points. [*Hint*: Your intuition should tell you that the points R_1, R_2, and R_3 are in the intersection of the planes determined by the triangles (P_1, P_2, P_3) and (Q_1, Q_2, Q_3). Prove this and use the dimension theorem to show that this intersection has dimension one.] [4.4]

8. Let $(\mathcal{P}, \Lambda)$ be an n-dimensional projective space, $A \in \mathcal{F}(\mathcal{P})$, p-dim $(A) = k$. Show that A itself is a k-dimensional projective space where the points are the points of $\mathcal{P}$ in A and the lines are the lines in Λ that are contained in A.

9. Let $(\mathcal{P}, \Lambda)$ be an n-dimensional projective space and H a hyperplane.
   (a) Show that $(\mathcal{P}/H)$ is a lattice.
   (The remaining parts of this exercise are the incidence relations as found in Chapter 2. The words points, lines, and so on, will refer to affine points, affine lines, and so on.)
   (b) The join of two distinct points is a line.
   (c) The join of three noncollinear points is a plane.
   (d) If A/H, $B/H \in \mathcal{F}(\mathcal{P}/H)$ we define parallelism by

   $$(A/H) \parallel (B/H) \Leftrightarrow (A \cap H \subseteq B \cap H \text{ or } B \cap H \subseteq A \cap H).$$

   Show that if P/H is an affine point and L/H is an affine line not containing P/H, there is a unique affine plane A/H containing both and a unique line M/H in A/H such that $(P/H) \subseteq (M/H)$ and $(L/H) \parallel (M/H)$.

SECTION 4.4
LATTICES AND DUALITY

In Section 4.3 we have shown that if $(\mathscr{P}, \Lambda)$ is an n-dimensional projective space then $\mathscr{F}(\mathscr{P})$ is a modular lattice of finite length with the property that every pair of elements of equal dimension have a common complement. In this section we shall show that every lattice of this type is the lattice of flats of a finite-dimensional projective space and we shall exploit this fact to develop the duality theory for projective geometry (refer to Chapter 2, Section 2.2, Exercises 6, 7, and 8 for lattice concepts). As a linguistic convenience we make the following definition.

DEFINITION 1 A modular lattice F of finite length with the property that every pair of elements of equal height have a common complement is called a *P-lattice*.

As standard notation from this point through the proof of Theorem 1, F will denote a P-lattice with "$+$" and "$\cap$" as operations, $\mathscr{P}$ the set of elements of F of height one which will be called points, and Λ the set of elements of height two which will be called lines. Our intent, as you might guess from the notation, is to show that $(\mathscr{P}, \Lambda)$ is a finite dimensional projective space and that F is $\mathscr{F}(\mathscr{P})$.

DEFINITION 2 Let $A, B \in F$; we say that B *covers* A if $A \subseteq B, A \neq B$ and for $C \in F$, $A \subseteq C \subseteq B \Rightarrow C = A$ or $C = B$.

Lemma 1

If $A_1, A_2, B \in F$, $A_1 \neq A_2$, and B covers both A_1 and A_2, then A_1 and A_2 cover $A_1 \cap A_2$.

Proof

(See Figure 4.7.) $A_1 \subseteq A_1 + A_2 \subseteq B$ and $A_2 \neq A_1 \Rightarrow A_1 \neq A_1 + A_2$ so $B = A_1 + A_2$. Clearly $A_1 \cap A_2 \subseteq A_1$. If $A_1 \cap A_2 = A_1$ then $A_1 \subseteq A_2 \subseteq B$ which is a contradiction since A_2 is distinct from both A_1 and B and B covers A_1. Now suppose we have $C \in F$ such that $A_1 \cap A_2 \subseteq C \subseteq A_1$. Since $C \subseteq A_1$, $C \cap A_2 \subseteq A_1 \cap A_2$ but $A_1 \cap A_2 \subseteq C$ and $A_1 \cap A_2 \subseteq A_2$ so $A_1 \cap A_2 \subseteq C \cap A_2$ which yields $C \cap A_2 = A_1 \cap A_2$. Also we have $A_2 \subseteq C + A_2 \subseteq A_1 + A_2 = B$ so $C + A_2 = A_2$ or $C + A_2 = A_1 + A_2$. If $C + A_2 = A_2$ then $C \subseteq A_2$ which

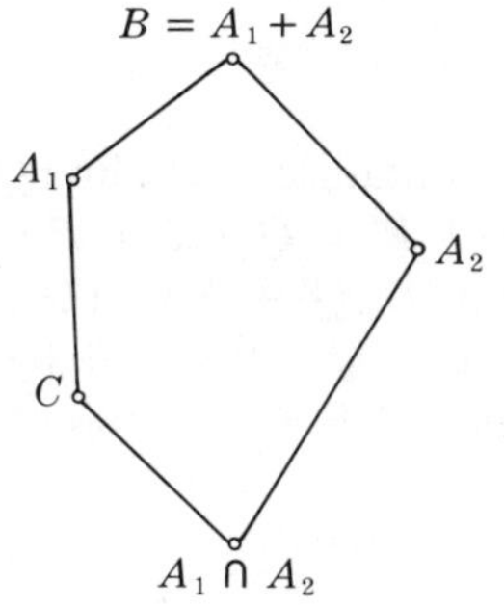

Figure 4.7

implies $C \subseteq A_1 \cap A_2$ and hence $C = A_1 \cap A_2$. If $C + A_2 = A_1 + A_2$ then

$$\begin{aligned} C &= C + (A_2 \cap C) = C + (A_2 \cap A_1), \\ &= (C + A_2) \cap A_1 \qquad \text{(by the modular law)}, \\ &= (A_1 + A_2) \cap A_1 = A_1. \end{aligned}$$

Therefore A_1 covers $A_1 \cap A_2$ and similarly A_2 covers $A_1 \cap A_2$.

Proposition 1

If P_1, P_2 are distinct elements of $\mathscr{P}$, then $P_1 + P_2$ is the unique element of Λ containing P_1 and P_2, that is, $(\mathscr{P}, \Lambda)$ satisfies PS (1).

Proof

We must first show that $P_1 + P_2 \in \Lambda$. As a first step we observe that $P_1 + P_2$ covers P_1 for suppose $A \in F, P_1 \subseteq A \subseteq P_1 + P_2$ and $A \neq P_1 + P_2$. Then $P_2 \subseteq A$ so $0 = A \cap P_2 = P_1 \cap P_2$ and, moreover, $P_1 + P_2 \subseteq A + P_2 \subseteq P_1 + P_2$ which implies $P_1 + P_2 = A + P_2$. Therefore, using the modular law, we have

$$P_1 = P_1 + (P_1 \cap P_2) = P_1 + (P_2 \cap A) = (P_1 + P_2) \cap A = A.$$

Let $0 \subseteq A_1 \subseteq \cdots \subseteq A_n = P_1 + P_2$ be a proper chain of maximal length from 0 to $P_1 + P_2$. Clearly $n \geq 2$ and $P_1 + P_2$ covers A_{n-1}. Using Lemma 1 we see that either $A_{n-1} = P_1$ and hence $h(P_1 + P_2) = 2$ or A_{n-1} and P_1 cover $A_{n-1} \cap P_1$. But P_1 being a point covers only 0, so A_{n-1} covers 0 and $h(P_1 + P_2) = 2$.

For the uniqueness, if $L \in \Lambda, P_1 \subseteq L$, and $P_2 \subseteq L$, then $P_1 + P_2 \subseteq L$ and, since $h(L) = h(P_1 + P_2)$, it follows that $P_1 + P_2 = L$.

Proposition 2

If P, Q, R, Q', R' are five distinct points, $P \notin Q + R$, $Q' \in P + Q$, and $R' \in P + R$, then $(Q' + R') \cap (Q + R) \in \mathscr{P}$ (that is, PS (2) holds).

Proof

$(Q' + R') \cap (Q + R)$ being contained in the line $Q + R$ is either a line, a point, or 0. If it is a line, then $Q' + R' = Q + R = R + R'$ but $P \in R + R'$ so $P \in Q + R$, which is a contradiction. If $(Q' + R') \cap (Q + R) = 0$ we have a contradiction to the modular law for $Q = Q + 0 = Q + ((Q' + R') \cap (Q + R))$ whereas $Q + R = (Q + (Q' + R')) \cap (Q + R)$. Therefore $(Q' + R') \cap (Q + R) \in \mathscr{P}$.

Proposition 3

Every element of Λ contains at least three points.

Proof

Let $L \in \Lambda$. Since $h(L) = 2$ we have a proper chain of maximal length $0 \subseteq P_1 \subseteq L$; P_1 covers 0 by the maximality, so P_1 is a point. Let P_1' be a complement of P_1 (the common complement of the pair P_1, P_1 if you wish). Then $L \cap P_1' = 0 \Rightarrow P_1 = P_1 + 0 = P_1 + (P_1' \cap L) = (P_1 + P_1') \cap L = L$, a contradiction, so $L \cap P_1' \neq 0$ and hence is a point, say P_2; for it certainly contains a point and if it contains a line, that line would be L but $L = L \cap P_1' \Rightarrow L \subseteq P_1' \Rightarrow P_1 \subseteq P_1'$. Then $P_1 \neq P_2$ and $L = P_1 + P_2$. The pair P_1, P_2 have a common complement H. We claim that $H \cap L$ is a point, and hence the third point on L. The technique is similar to that used in the proof of Proposition 2. Since L is a line, $H \cap L$ is either a line, a point, or 0. If $H \cap L$ is a line then $H \cap L = L \Rightarrow L \subseteq H \Rightarrow P_1 \subseteq H$ and $P_2 \subseteq H$, contradicting the choice of H as a complement to P_1 and P_2. If $H \cap L = 0$, then

$$P_1 = P_1 + 0 = P_1 + (H \cap L) = (P_1 + H) \cap L = I \cap L = L,$$

which is a contradiction so $H \cap L \in \mathscr{P}$.

Proposition 4

If F is of length $n + 1$, then there exists a basis of $n + 1$ points for $\mathscr{P}$.

Proof

If F is of length $n + 1$ this means $h(I) = n + 1$ so there is a proper chain of maximal length

$$0 = A_0 \subseteq A_1 \subseteq \cdots \subseteq A_n \subseteq A_{n+1} = I.$$

By the same argument used in Proposition 3 we can show that $P_i = A_i' \cap A_{i+1}$, for $i = 0, 1, \cdots, n$ (where A_i' is a complement to A_i) is a point. Then $A_{i+1} = P_0 + \cdots + P_i$ for all i whence $I = P_0 + \cdots + P_n$. The set $\{P_0, P_1, \cdots, P_n\}$ is independent, for if not there is a smallest integer k such that $\{P_0, P_1, \cdots, P_k\}$ is dependent, which would imply $P_k \in P_0 + \cdots + P_{k-1}$ (Proposition 3, Section 4.3) and hence $A_{k+1} = A_k$. If $P \in \mathscr{P}$ then $P \subseteq I \Rightarrow P \subseteq P_0 + \cdots + P_n$ so $\mathscr{P} = P_0 + \cdots + P_n$ and we can identify $\mathscr{P}$ and I.

As a remark we might add that $0 \in F$ can be identified with the empty flat $\varnothing$. We can now summarize and extend these results with the following theorem.

Theorem 1

Let F be a P-lattice of length $n + 1$,

$$\mathscr{P} = \{P \in F : h(P) = 1\}, \Lambda = \{L \in F : h(L) = 2\}.$$

Then $(\mathscr{P}, \Lambda)$ is an n-dimensional projective space and $\mathscr{F}(\mathscr{P})$ is F.

Proof

Propositions 1–4 assert that $(\mathscr{P}, \Lambda)$ satisfies axioms PS (1)–PS (4). Every element of $\mathscr{F}(\mathscr{P})$ can be written as the sum of points, but F is a lattice and so closed under "+", which says that every flat is an element of F. Conversely every element of F can be written as a sum of points, using the argument of Proposition 4 where we showed that I is the sum of points, and hence every element of F is in $\mathscr{F}(\mathscr{P})$.

We have thus established the fact that the study of projective geometry is coextensive with the study of a certain class of lattices. We shall now exploit this via duality.

DEFINITION 3 Let F be a lattice (for the moment we do not need to assume F is a P-lattice). A *dual to F* is a lattice F^* together with a 1-1 onto map $\delta : F \to F^*$ such that for $A, B \in F$, $\delta(A + B) = \delta(A) \cap \delta(B)$ and $\delta(A \cap B) = \delta(A) + \delta(B)$.

Proposition 5

Let F be a lattice. Then

(a) there exists a dual $\delta : F \to F^*$ of F;
(b) if $\delta : F \to F^*$ is a dual to F, then $\delta^{-1} : F^* \to F$ is a dual to F^*;
(c) if $\delta : F \to F^*$ and $\gamma : F \to G$ are two duals of F, then there is a lattice isomorphism $\theta : F^* \to G$.

Proof

(a) Let $F^* = \{A^* : A \in F\}$, that is, for each element $A \in F$ we take a symbol A^* and let F^* be the set of these symbols. We then have a 1-1 onto map $\delta : F \to F^*$ defined by $\delta(A) = A^*$. We now make F^* into a lattice in such a way that our map has the properties we want. This is much like the way we defined the addition and multiplication of matrices. If $A^*, B^* \in F^*$ we define $A^* + B^* = (A \cap B)^*$ and $A^* \cap B^* = (A + B)^*$. We leave it as an exercise to check that F^* is a lattice. Clearly $\delta(A + B) = (A + B)^* = A^* \cap B^* = \delta(A) \cap \delta(B)$ and $\delta(A \cap B) = (A \cap B)^* = A^* + B^* = \delta(A) + \delta(B)$.

(b) Since δ is 1-1 and onto, δ^{-1} is a map. Noting that $\delta^{-1}(A^*) = A$ we have

$$\begin{aligned}\delta\delta^{-1}(A^* + B^*) &= A^* + B^* = (A \cap B)^* = \delta(A \cap B)\\ &= \delta(\delta^{-1}(A^*) \cap \delta^{-1}(B^*)),\end{aligned}$$

and since δ is 1-1,

$$\delta^{-1}(A^* + B^*) = \delta^{-1}(A^*) \cap \delta^{-1}(B^*).$$

Similarly $\delta^{-1}(A^* \cap B^*) = \delta^{-1}(A^*) + \delta^{-1}(B^*)$.

(c) Let $\theta = \gamma\delta^{-1}$. θ is 1-1 and onto since it is the composite of 1-1 onto maps. If $A^*, B^* \in F^*$ then

$$\begin{aligned}\theta(A^* + B^*) &= \gamma\delta^{-1}(A^* + B^*)\\ &= \gamma(A \cap B)\\ &= \gamma(A) + \gamma(B)\\ &= \gamma\delta^{-1}(A^*) + \gamma\delta^{-1}(B^*)\\ &= \theta(A^*) + \theta(B^*).\end{aligned}$$

Similarly, $\theta(A^* \cap B^*) = \theta(A^*) \cap \theta(B^*)$.

Theorem 2

Let $(\mathscr{P}, \Lambda)$ be an n-dimensional projective space. Then there exists an n-dimensional projective space $(\mathscr{P}^*, \Lambda^*)$ (called a *dual to* $(\mathscr{P}, \Lambda)$) such that $\mathscr{F}(\mathscr{P}^*)$ is a dual lattice to $\mathscr{F}(\mathscr{P})$.

Proof

The lattice $\mathscr{F}(\mathscr{P})$ has a dual $\delta: \mathscr{F}(\mathscr{P}) \to \mathscr{F}(\mathscr{P})^*$. If we can show that the dual of a P-lattice is a P-lattice then we let $(\mathscr{P}^*, \Lambda^*)$ be the projective space, whose existence Theorem 1 guarantees, such that $\mathscr{F}(\mathscr{P}^*) = \mathscr{F}(\mathscr{P})^*$.

Note first that $A \subseteq B \Leftrightarrow A = A \cap B \Leftrightarrow B = A + B$, so $A \subseteq B \Leftrightarrow \delta(A) \supseteq \delta(B)$. In particular $\delta(\mathscr{P})$ is the $\varnothing$-element of $\mathscr{F}(\mathscr{P})^*$ and $\delta(\varnothing) = \mathscr{P}^*$. Then if $A \in \mathscr{F}(\mathscr{P})$ and $A = A_0 \subseteq A_1 \subseteq \cdots \subseteq A_k = \mathscr{P}$ is a proper chain of maximal length from A to $\mathscr{P}$, $\varnothing = \delta(A_k) \subseteq \delta(A_{k-1}) \subseteq \cdots \subseteq \delta(A_0) = \delta(A)$ is a maximal chain from $\varnothing$ to $\delta(A)$ in $\mathscr{F}(\mathscr{P})^*$. So, $h(\delta(A)) = n + 1 - h(A)$. If A' is a complement to A in $\mathscr{P}$, $\varnothing = A \cap A'$, and $\mathscr{P} = A + A' \Rightarrow \delta(A) + \delta(A') = \delta(\varnothing)$ and $\delta(A) \cap \delta(A') = \delta(\mathscr{P})$; so $\delta(A')$ is a complement to $\delta(A)$. If $h(A) = h(B)$ and C is their common complement $\delta(C)$ is the common complement of $\delta(A)$ and $\delta(B)$.

If $\delta(A) \subseteq \delta(C)$, then $C \subseteq A$ and

$$\begin{aligned}\delta(A) + (\delta(B) \cap \delta(C)) &= \delta(A) + \delta(B + C)\\ &= \delta(A \cap (B + C))\\ &= \delta((A \cap B) + C) \text{ (since } \mathscr{F}(\mathscr{P}) \text{ is modular)}\\ &= \delta(A \cap B) \cap \delta(C)\\ &= (\delta(A) + \delta(B)) \cap \delta(C),\end{aligned}$$

so $\mathscr{F}(\mathscr{P})^*$ is a P-lattice.

As a corollary to the proof, if not the theorem, we record the following useful fact.

Corollary

If $(\mathscr{P}, \Lambda)$ is an n-dimensional projective space and if $A \in \mathscr{F}(\mathscr{P})$, p-dim $(A) = k$, then p-dim $(\delta(A)) = n - (k + 1)$.

We shall now consider some applications of the duality theorem, beginning with a rather simple case. Let $(\mathscr{P}, \Lambda)$ be an n-dimensional projective space and A and B distinct flats of projective dimension $n - 2$ each containing a flat C of dimension $n - 3$. Using part (b) of Proposition 5 and Theorem 2 we know that $(\mathscr{P}, \Lambda)$ is the dual of a projective space $(\mathscr{P}', \Lambda')$. The pre-images of A and B, say A' and B', are lines and the pre-image of C, say C', is a plane. $A' \neq B'$ and $A' \subseteq C'$, $B' \subseteq C'$. Therefore $A' \cap B'$ is a point, which implies $A + B$ is a hyperplane. This fact is of course immediate from the observation that $C = A \cap B$ and from Proposition 6, Section 4.3.

A more significant topic to discuss from the point of view of duality is Desargues' theorem. Since Desargues' theorem is about triangles we begin by looking at the notion of a triangle. By definition, a triangle is a set of three noncollinear points denoted $\Delta = (P_1, P_2, P_3)$. If p-dim $(\mathscr{P}) = 2$ and $\delta: \mathscr{P} \to \mathscr{P}^*$ is a dual then $\delta(P_1)$, $\delta(P_2)$, $\delta(P_3)$ are lines which do not intersect in a point or alternatively the intersection of no pair is contained in the third. We shall define a *trilateral* in an arbitrary projective space to be a set of three distinct coplanar lines (L_1, L_2, L_3) that do not intersect in a point. Although, as we have seen, in a plane the dual of a triangle is a trilateral, in spaces of higher dimension this statement is no longer true and must be amended to say that the dual of a triangle is a set of three hyperplanes such that the intersection of no pair is contained in the third.

In a projective space of any dimension greater than or equal to 2 there is a 1-1 correspondence between triangles and trilaterals given by

$$(P_1, P_2, P_3) \to (P_2 + P_3, P_1 + P_3, P_1 + P_2)$$

and

$$(L_1, L_2, L_3) \to (L_2 \cap L_3, L_1 \cap L_3, L_1 \cap L_2).$$

The image of a triangle under this correspondence is called its *associated trilateral* and the image of a trilateral will be called its *associated triangle.*

DEFINITION 1 Let $\mathscr{P}$, $\mathscr{P}'$ be (finite dimensional) projective spaces. A 1-1 function α from $\mathscr{P}$ onto $\mathscr{P}'$ is called a *collineation* if

$$\alpha(P + Q) = \alpha(P) + \alpha(Q), \forall P, Q \in \mathscr{P}.$$

The condition $\alpha(P + Q) = \alpha(P) + \alpha(Q)$ bears deceptive resemblance to its analogue for linear maps on a vector space. The reader is cautioned that $\alpha(P + Q)$ denotes a *collection of points*, namely

$$\{\alpha(R): R \in P + Q\}.$$

We are using here, and throughout this section, this common abuse of functional notation for brevity. Thus our definition asserts that a collineation sends each line in $\mathscr{P}$ onto a line in $\mathscr{P}'$. Exercise 1 shows that with modest restrictions the apparently weaker condition

$$\alpha(P + Q) \subseteq \alpha(P) + \alpha(Q)$$

would yield the same results. Note that if $\mathscr{P}$ is a projective space of p-dimension 1, then any permutation of the points of $\mathscr{P}$ satisfies the requirements for being a collineation. For this reason we assume until further notice that all projective spaces have p-dimension greater than or equal to 2.

Proposition 1

Let $\alpha:\mathscr{P} \to \mathscr{P}'$ be a collineation. If $A \in \mathscr{F}(\mathscr{P})$, then $\alpha(A) \in \mathscr{F}(\mathscr{P}')$.

Proof

Let $\alpha(P), \alpha(Q) \in \alpha(A)$, for $P, Q \in A$. Then $\alpha(P) + \alpha(Q) = \alpha(P + Q) \subseteq \alpha(A)$.

Proposition 2

Let $\alpha:\mathscr{P} \to \mathscr{P}'$ be a collineation. Then $A, B \in \mathscr{F}(\mathscr{P}) \Rightarrow \alpha(A + B) = \alpha(A) + \alpha(B)$.

Proof

Let $\alpha(P) \in \alpha(A + B)$ for $P \in A + B$. Then $\exists R \in A$, $S \in B$ with $P \in R + S$. Hence $\alpha(P) \in \alpha(R + S) = \alpha(R) + \alpha(S) \subseteq \alpha(A) + \alpha(B)$, and $\alpha(A + B) \subseteq \alpha(A) + \alpha(B)$. Now let $P \in \alpha(A) + \alpha(B)$. Then $P \in \alpha(R) + \alpha(S)$ for $R \in A$, $S \in B$. Hence $P \in \alpha(R) + \alpha(S) = \alpha(R + S) \subseteq \alpha(A + B)$, establishing $\alpha(A) + \alpha(B) \subseteq \alpha(A + B)$.

Corollary

$P_1, \cdots, P_k \in \mathscr{P} \Rightarrow \alpha(P_1 + \cdots + P_k) = \alpha(P_1) + \cdots + \alpha(P_k)$.

The preceding construction is significant. In the special case where p-dim $H = 2$, we see that projective geometry does not allow the pathology reported in Section 4.1 concerning affine geometry. In general, if we keep the above notation and choose $R' \neq R$, $R' \notin H$, $R' \notin H'$, then we can construct a collineation $\alpha' : H' \to H$. The composite $\alpha'\alpha : H \to H$ is then a collineation, on the projective space H, which leaves fixed all points on a hyperplane $H \cap H'$ of H. It is precisely this kind of map that displays the intimate relationship between Desargues' theorem and the existence of sufficiently many collineations to coordinatize a projective space. Let us record some basic facts concerning these maps.

DEFINITION 2 A collineation $\alpha : \mathscr{P} \to \mathscr{P}$ is called a *perspectivity* if there is a hyperplane H of $\mathscr{P}$ such that $\alpha(S) = S$, $\forall S \in H$. The hyperplane H is called the *axis* of the perspectivity.

Theorem 3

Let $\alpha : \mathscr{P} \to \mathscr{P}$ be a perspectivity with axis H. If $\alpha(P) = P$ and $\alpha(Q) = Q$ for distinct points P, $Q \notin H$, then α is the identity map on $\mathscr{P}$.

Proof

(See Figure 4.9.) Let $R \in \mathscr{P}$, $R \notin H$, $R \notin P + Q$. Then $(P + R) \cap H = R'$ and

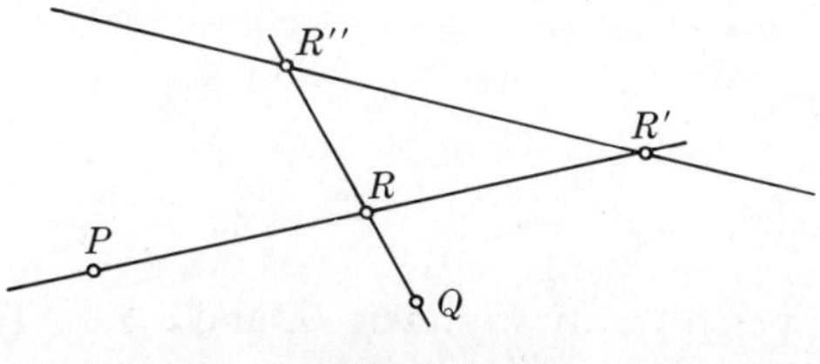

Figure 4.9

$(Q + R) \cap H = R''$ are fixed points under the map α since they belong to H. Thus $R \in P + R' \Rightarrow \alpha(R) \in \alpha(P) + \alpha(R') = P + R'$, and $R \in Q + R'' \Rightarrow \alpha(R) \in \alpha(Q) + \alpha(R'') = Q + R''$. Hence $\alpha(R) = (P + R') \cap (Q + R'') = R$. Now let T be any point of $\mathscr{P}$ not in H. Then for at least one pair (P, Q), (P, R), (Q, R), we know T is not in their sum. Using any such pair, the above argument shows $\alpha(T) = T$. Hence α is the identity.

Corollary

Let $\alpha : \mathscr{P} \to \mathscr{P}$ be a perspectivity with axis H. Then α is completely determined by its action on two distinct points not in H.

Proof

Suppose $\beta: \mathscr{P} \to \mathscr{P}$ is also a perspectivity with axis H. If for $P, Q \notin H, P \neq Q$ we have $\alpha(P) = \beta(P)$ and $\alpha(Q) = \beta(Q)$, then $\beta^{-1}\alpha(P) = P$ and $\beta^{-1}\alpha(Q) = Q$, so that $\beta^{-1}\alpha =$ the identity map on $\mathscr{P}$. Hence $\alpha = \beta$.

Actually this corollary can be replaced by a sharper result. Suppose $\alpha: \mathscr{P} \to \mathscr{P}$ is a perspectivity with axis H and that $\alpha(P_0) = P_0$ for some $P_0 \notin H$. Then every line through P_0 is mapped onto itself by α. For if $P_0 \in L$, we let $Q_0 = L \cap H$. Then

$$\alpha(L) = \alpha(P_0 + Q_0) = \alpha(P_0) + \alpha(Q_0) = P_0 + Q_0 = L.$$

On the other hand if $\alpha(P) \neq P$, $\forall P \in H$ (the only other possibility), we let $Q \notin H$. Then the line $Q + \alpha(Q)$ is mapped onto itself by α, since, putting $Q_0 = (Q + \alpha(Q)) \cap H$, we have

$$\begin{aligned}\alpha(Q + \alpha(Q)) &= \alpha(Q + Q_0) = \alpha(Q) + \alpha(Q_0) = \alpha(Q) + Q_0 \\ &= \alpha(Q) + Q.\end{aligned}$$

With these facts in mind we assert:

Lemma

Let $\alpha: \mathscr{P} \to \mathscr{P}$ be a collineation. If L, L' are distinct lines in $\mathscr{P}$ such that $L \cap L' = P_0$, and if $\alpha(L) = L$ and $\alpha(L') = L'$, then $\alpha(P_0) = P_0$.

Proof

The proof is obvious.

Theorem 4

Let $\alpha: \mathscr{P} \to \mathscr{P}$ be a perspectivity with axis H, and $\alpha \neq 1$. Then there exists a unique point $P_0 \in \mathscr{P}$ such that $\alpha(L) = L$ for each line L containing P_0. We call P_0 the *center* of α.

Proof

Uniqueness: Suppose P_0, P_1 are distinct points satisfying the theorem. Let $Q \in \mathscr{P}$, $Q \notin P_0 + P_1$. Then the lines $P_0 + Q$, $P_1 + Q$ are mapped onto themselves by α, so, by the Lemma, $\alpha(Q) = Q$. Since there exist at least two points Q, Q' neither in H nor in $P_0 + P_1$ (why?) we conclude $\alpha = 1$, contrary to hypothesis. Note that it did not matter whether P_0, P_1 belong to H.

Existence: By the remarks immediately preceding the Lemma, if $\alpha(P_0) = P_0$ for some $P_0 \notin H$, then P_0 is a (and hence *the*) center of α. It remains only to show that if α leaves fixed no point outside of H, then α has a center in H. So suppose α is such a map. Let $Q \notin H$, and let $(Q + \alpha(Q)) \cap H = P_0$. We know that $Q + \alpha(Q)$ is left fixed by α. If for some $P \notin Q + P_0$ we had $(P + \alpha(P)) \cap H \neq P_0$, then $(P + \alpha(P)) \cap (Q + \alpha(Q))$ is a fixed point of α not in H. Thus P_0 is the center of α.

DEFINITION 3 Let $\alpha: \mathscr{P} \to \mathscr{P}$ be a perspectivity with axis H and center P_0. If $P_0 \in H$ we call α an *elation*; if $P_0 \notin H$ we call α a *homology*.

Corollary

Let $\alpha: \mathscr{P} \to \mathscr{P}$ be an elation with axis H. Then α is completely determined by the image of one point $P \notin H$.

The following facts will be of considerable use to us in later developments. Their proofs are left to the reader (Exercise 3).

Theorem 5

Let H be a hyperplane of $\mathscr{P}$. Then

(a) the set Aut $(\mathscr{P})$ of all collineations is a group under composition;
(b) the set of all perspectivities with axis H is a subgroup of Aut $(\mathscr{P})$;
(c) the set of elations with axis H is a subgroup of Aut $(\mathscr{P})$;
(d) if α_1, α_2 are perspectivities with axis H and center P (whether $P \in H$ or $P \notin H$), then ${\alpha_1}^{-1}$ has axis H and center P, and $\alpha_1\alpha_2$ has axis H and center P, or else $\alpha_1\alpha_2 = 1$;
(e) if α is a homology with axis H and center P, and if γ is an elation with axis H and center P_0, then $\alpha\gamma\alpha^{-1}$ is an elation with axis H and center P_0.

Now that we know a good deal about perspectivities, let us show how some of them are constructed. Our first result in this direction uses the ideas of the example given after Theorem 2 to prove a promised result.

Theorem 6 (Desargues)

Let p-dim $(\mathscr{P}) \geq 3$. Let (P_1, P_2, P_3), (Q_1, Q_2, Q_3) be triangles with six distinct vertices and six distinct sides *all lying in a plane H*. If

$$(P_1 + Q_1) \cap (P_2 + Q_2) \cap (P_3 + Q_3) = P,$$

then the points $R_1 = (P_2 + P_3) \cap (Q_2 + Q_3)$, $R_2 = (P_1 + P_3) \cap (Q_1 + Q_3)$, $R_3 = (P_1 + P_2) \cap (Q_1 + Q_2)$ are collinear.

Proof

Let $L = R_2 + R_3$. Let H' be a second plane in $\mathscr{P}$ such that $H \cap H' = L$. (For example, let $H' = L + S$ for any point $S \notin H$.) Choose $R \in H + H'$, such that $R \notin H$, $R \notin H'$. (Why is this possible?) Let $\alpha_1 : H \to H'$ be the collineation given by

$$\alpha_1(Q) = (Q + R) \cap H', \forall Q \in H.$$

Let $R' = (P + R) \cap (\alpha_1(P_1) + Q_1)$. (See Figures 4.10 and 4.11.) Now define

$$\alpha_2 : H' \to H \text{ by } \alpha_2(Q) = (Q + R') \cap H.$$

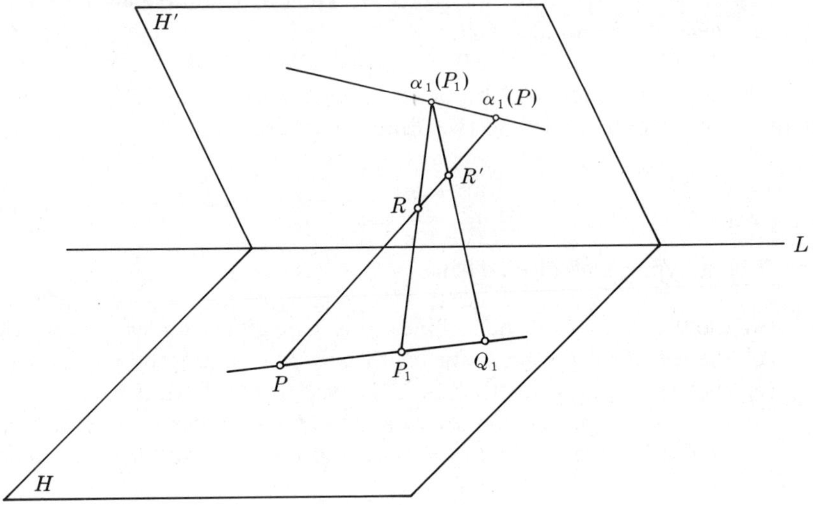

Figure 4.10

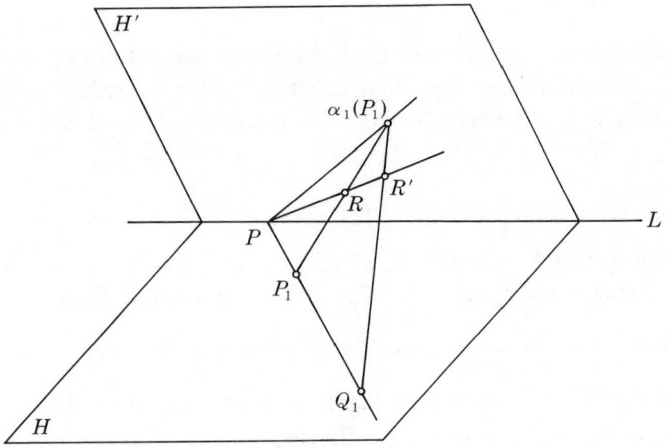

Figure 4.11

In order that α_2 be a collineation, it is necessary that R' is indeed a *point* and $R' \notin H$, $R' \notin H'$. But the lines $P + R$ and $\alpha_1(P_1) + Q_1$ cannot be equal since that would say $R \in P + Q_1 \subseteq H$. Moreover both of the lines $P + R$ and $\alpha_1(P_1) + Q_1$ lie in the plane $(P + P_1) + (\alpha(P) + \alpha(P_1))$ determined by the line $P + P_1$ and its image under the collineation α_1. Finally the point R' cannot be in H_1 or H_2, since that would imply that R would be in H_1 or H_2, because $R \in P + R'$. Thus α_2 is a collineation. The composite $\alpha = \alpha_2 \alpha_1 : H \to H$ is therefore a collineation that leaves L fixed, that is, α is a perspectivity with axis L. In addition $\alpha(P) = P$ and $\alpha(P_1) = Q_1$; hence $\alpha(P + P_1) = P + P_1$, and P is therefore the center of α. Whether or not $P \in L$ is immaterial.

Now look at Figure 4.12. We know that $\alpha(R_2) = R_2$ and $\alpha(R_3) = R_3$ since $R_2, R_3 \in L$. Thus

$$\begin{aligned}\alpha(P_2) \in \alpha(P_1 + R_3) \cap \alpha(P + P_2) &= (Q_1 + R_3) \cap (P + P_2)\\ &= Q_2 .\end{aligned}$$

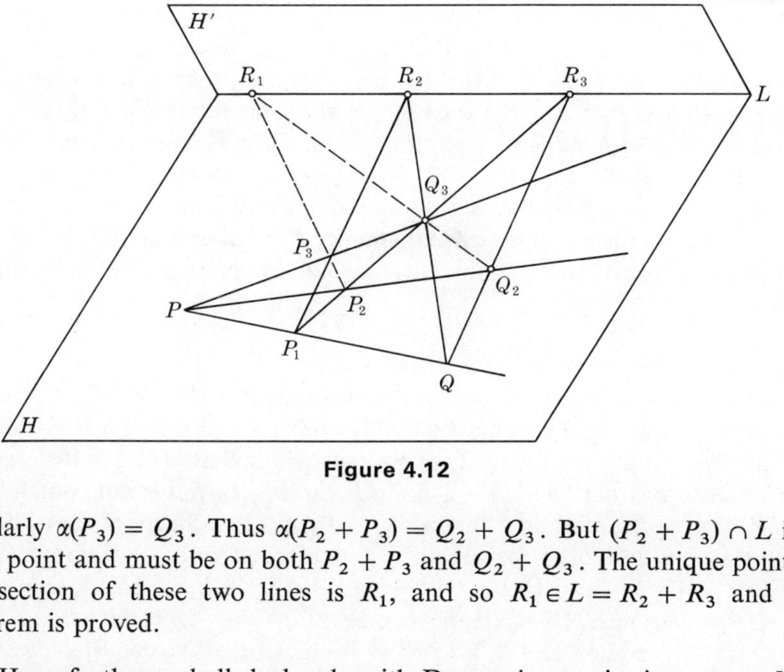

Figure 4.12

Similarly $\alpha(P_3) = Q_3$. Thus $\alpha(P_2 + P_3) = Q_2 + Q_3$. But $(P_2 + P_3) \cap L$ is a fixed point and must be on both $P_2 + P_3$ and $Q_2 + Q_3$. The unique point of intersection of these two lines is R_1, and so $R_1 \in L = R_2 + R_3$ and the theorem is proved.

Henceforth we shall deal only with Desarguian projective spaces. It is the presence of this theorem which enables us to construct all possible perspectivities in $\mathscr{P}$. More precisely we have:

Theorem 7

Let $\mathscr{P}$ be a Desarguian projective space, with p-dim $(\mathscr{P}) \geq 2$. Let H be a hyperplane of $\mathscr{P}$ and let P, P_1, Q_1 be distinct collinear points such that $P_1, Q_1 \notin H$. Then there exists a unique perspectivity $\alpha : \mathscr{P} \to \mathscr{P}$ with axis H and center P such that $\alpha(P_1) = Q_1$.

Proof

Let P_2 be any point not in H and not on $P + P_1$. Let $R_3 = (P_1 + P_2) \cap H$, and $Q_2 = (Q_1 + R_3) \cap (P + P_2)$. First note that R_3 is a single point since H is a hyperplane and $(P_1 + P_2)$ is a line not in H. Also Q_2 is a single point since $(Q_1 + R_3)$ and $(P + P_2)$ are distinct (why?) lines lying in the plane $(P + P_1 + P_2)$. (See Figure 4.12.)

If P_3 is any point not in H and not in $P + P_1$, and not in $P + P_2$, then we put $R_2 = (P_1 + P_3) \cap H$ and $Q_3 = (Q_1 + R_2) \cap (P + P_3)$. Again, R_2 and Q_3 are single points. By Desargues' theorem, the point $R_1 = (P_2 + P_3) \cap (Q_2 + Q_3)$ is on the line $R_2 + R_3$. In particular, $R_1 \in H$. Thus Q_3 can be realized by $Q_3 = (Q_1 + R_1) \cap (P + P_3)$ where $R_1 = (P_2 + P_3) \cap H$.

We define $\alpha: \mathscr{P} \to \mathscr{P}$ as follows. We are given P, P_1, Q_1; we select P_2 and determine Q_2 as above. Now for any point $P_3 \notin H$, $P_3 \neq P$, we must have $P_3 \notin P + P_1$ or $P_3 \notin P + P_2$ (or both). If the former holds, we *define* $\alpha(P_3) = Q_3$, where as above, $Q_3 = (Q_1 + R_2) \cap (P + P_3)$ and $R_2 = (P_1 + P_3) \cap H$. If $P_3 \notin P + P_2$, we define $\alpha(P_3) = Q_3'$ where $Q_3' = (Q_1 + R_1) \cap (P + P_3)$ and $R_1 = (P_2 + P_3) \cap H$. The second paragraph shows that if $P_3 \notin P + P_1$ and $P_3 \notin P + P_2$, then we get the same result $Q_3' = Q_3$. In other words α is *well-defined* for all points $P_3 \notin H$, $P_3 \neq P$. We complete the definition by stipulating that $\alpha(P) = P$, and $\alpha(S) = S$, $\forall S \in H$.

Before showing that α is the desired perspectivity we make one final observation: In the above construction, once P_3 is chosen and Q_3 determined, we can use (P_3, Q_3) to determine the image of any point $P_4 \notin P + P_3$, that is,

$$\alpha(P_4) = (P + P_4) \cap (Q_3 + R_0) \qquad (*)$$

where $R_0 = (P + P_4) \cap H$. (See Exercise 2.)

It remains to show that α is a collineation. It will follow from the construction of α and the Corollaries to Theorems 3 and 4 that α is the desired perspectivity and it is unique. Thus we must show that $\alpha(L) \in \Lambda$ for each line L of $\mathscr{P}$. By construction $\alpha(L) = L$ if $P \in L$. So let L be a line not containing P. Let $P_0 = L \cap H$. We know that $\alpha(P_0) = P_0$, so it suffices to show that the images of any two other points on L are collinear with P_0. Choose $R_1 \in L$, $R_1 \neq P_0$. Then (see Figure 4.13) $\alpha(R_1) = (P + R_1) \cap (S_0 + Q_1)$ where $S_0 = (P_1 + R_1) \cap H$. (If R_1 happens to equal P_1, then omit this step and take P_1, Q_1 for R_1, $\alpha(R_1)$.) Let $R_2 \in L$, $R_2 \neq R_1$, $R_2 \neq P_0$. We use R_1 and $\alpha(R_1)$ to compute $\alpha(R_2)$: $\alpha(R_2) = (P + R_2) \cap (S_0' + \alpha(R_2))$ where $S_0' = (R_1 + R_2) \cap H$. But then $S_0' = P_0$ and we get $\alpha(R_2) \in P_0 + \alpha(R_1)$ as desired. Now we may either invoke Exercise 1 or observe (see Figure 4.13) that α maps L *onto* $P_0 + \alpha(R_1)$. Q.E.D.

Summary Let $\mathscr{P}$ be a Desarguian space, p-dim$(\mathscr{P}) \geq 2$, and let H be a hyperplane of $\mathscr{P}$. Any collineation from $\mathscr{P}$ onto $\mathscr{P}$ that leaves the points of H fixed is a perspectivity, that is, either an elation or a homology. All possible elations

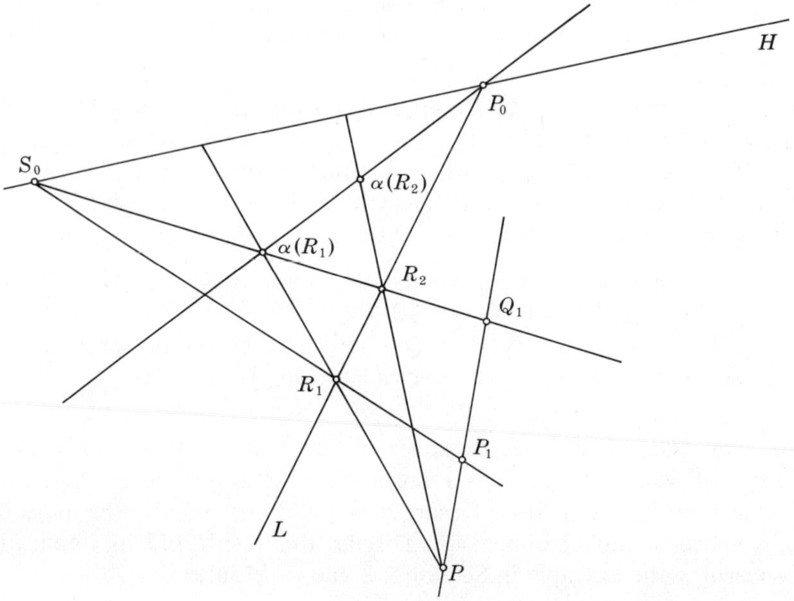

Figure 4.13

and homologies exist, and each is completely determined by its center P and the image of one point P_1 other than P not in H. Conversely (examine the proof of Theorem 5) *any* projective space in which all possible perspectivities exist must be Desarguian.

EXERCISES

1. Let $\alpha:\mathscr{P} \to \mathscr{P}$ be a 1-1 map of a projective space $\mathscr{P}$ onto itself such that $\alpha(P + Q) \subseteq \alpha(P) + \alpha(Q)$, $\forall P, Q \in \mathscr{P}$. Show that α is a collineation.

2. Verify the assertion (*) in the proof of Theorem 6.

3. Prove Theorem 5.

*4. Let H be a hyperplane of $\mathscr{P}$ and let α be a perspectivity of $\mathscr{P}$ with axis H. Let $\bar{\alpha}$ be the restriction of α to $\mathscr{P}/H$. Then:

   (a) show $\bar{\alpha}$ is 1-1 and onto and sends each line onto a parallel line;
   (b) show that if α is an elation, then $Q + \alpha(Q) \parallel R + \alpha(R)$, $\forall Q, R \in \mathscr{P}/H$. [4.6]

5. Let $\mathscr{P} = \mathscr{P}(\mathbb{Z}_2{}^3)$ in Theorem 5. Determine the number of elements in Aut ($\mathscr{P}$), as well as in some of the subgroups mentioned.

6. Prove that the converse to Desargues' theorem holds in any Desarguian space of projective dimension $n \geq 2$. [*Hint*: You are given triangles (P_1, P_2, P_3), (Q_1, Q_2, Q_3) with six distinct vertices and six distinct sides such that $R_1 + R_2 + R_3$ is a line where $R_1 = (P_2 + P_3) \cap (Q_2 + Q_3)$, and so on. The problem is to show that $\bigcap_{i=1}^{3} (P_i + Q_i)$ is a point.
   (a) Show that there is a hyperplane H containing $R_1 + R_2 + R_3$ but not containing P_i or Q_i, $i = 1, 2, 3$.
   (b) Let $P = (P_1 + Q_1) \cap (P_2 + Q_2)$, and let α be the perspectivity with axis H and center P such that $\alpha(P_1) = Q_1$.]

7. An *imbedding* of a projective space $\mathscr{P}$ into a space $\mathscr{P}'$ is a collineation $\gamma: \mathscr{P} \to \mathscr{P}''$ whose image $\mathscr{P}''$ is a flat in $\mathscr{P}'$. A consequence of this section is that a projective plane is Desarguian if and only if it can be imbedded in a space of higher dimension. Discuss this result and its implication concerning the example in Section 4.2 due to Moulton.

SECTION 4.6
COORDINATIZATION

Let $\mathscr{P}$ be a Desarguian projective space of dimension $n \geq 2$, and let H be a hyperplane of $\mathscr{P}$. Intuitively, the associated affine space $\mathscr{P}/H$ corresponds to an n-dimensional vector space over some division ring. We would like to show that in fact there does exist a division ring D and a map $\alpha: \mathscr{P}/H \to D^n$ that is 1-1 and onto and sends lines onto lines. Once we have this, we can show that $\mathscr{P}$ and $\mathscr{P}(D^{n+1})$ are *projectively equivalent*, that is, there exists a collineation $\varphi: \mathscr{P} \to \mathscr{P}(D^{n+1})$ (denoted $\mathscr{P} \approx \mathscr{P}(D^{n+1})$).

At our disposal are certain facts about the points and lines of $\mathscr{P}/H$ resulting from the corresponding facts about points and lines in $\mathscr{P}$. We also have available certain maps of $\mathscr{P}/H$ onto itself resulting from the restrictions to P/H of the perspectivities of $\mathscr{P}$ with axis H. The existence of these maps will allow us to construct the division ring D, with which we shall coordinatize $\mathscr{P}/H$. To gain insight into this construction let us return briefly to the context of a vector space over a division ring to observe how the elements of the division ring and of the vector space are related to certain maps which preserve the affine structure of the space. The reader is encouraged to reread Sections 2.5–2.7 of Chapter 2 and to familiarize himself with the exercises for that material.

Let ${}_DV$ be a vector space. For each $\alpha \in D$ we define a map by $\bar{\alpha}:V \to V$

$$\bar{\alpha}(x) = \alpha x, \forall x \in V.$$

Put $\bar{D} = \{\bar{\alpha}:\alpha \in D\}$, which we call the set of *dilatations* (about the origin). In particular, $\bar{0}$ collapses V onto O, and $\bar{1}$ is the identity map. Even when D is not commutative, the map $\bar{\alpha}$ is (except for $\bar{0}$) a 1-1 map of V onto V leaving O and no other point fixed (except for $\bar{1}$) and sends each affine line L of V onto a line L' such that $L' \parallel L$. Every line through O is mapped onto itself. In short, these dilatations behave precisely like the restrictions to $\mathscr{P}/H$ of the homologies with axis H and fixed center $P \notin H$ (see Section 4.5, Exercise 4). Intuitively, $\bar{\alpha}$ stretches V uniformly about O. Algebraically, $\bar{\alpha}$ is a semilinear map whose associated automorphism is $\beta \to \alpha\beta\alpha^{-1}$.

The second class of maps on V which are of present interest is the set T of all translations. For each $x \in V$ we have the translation $t_x:V \to V$ given by

$$t_x(y) = x + y, \forall y \in V.$$

If $x = O$, then $t_x = 1$. Otherwise, t_x is a 1-1 map of V onto V leaving no point fixed and sending each line L of V onto a line L' such that $L' \parallel L$. Any line containing both a point and its image under t_x is mapped onto itself. Thus the translations behave precisely like the restrictions to $\mathscr{P}/H$ of the elations of $\mathscr{P}$ with axis H. A translation, like an elation, is completely determined by the image of one point.

Thus there exists a 1-1 correspondence between the scalars D and the dilatations $\bar{D}$ as well as a 1-1 correspondence between the vectors V and the translations T. Moreover, in a fairly natural way, the algebraic structure of D and V can be carried over via these correspondences to impose algebraic structure on $\bar{D}$ and T. Specifically, the product (that is, composite) $\bar{\alpha} \cdot \bar{\beta}$ of two dilatations is the dilatation which maps the vector x to $(\alpha\beta)x$, that is, $\bar{\alpha} \cdot \bar{\beta} = \overline{(\alpha\beta)}$. Likewise the composite $t_x t_y$ of two translations is that translation that maps the vector O to $x + y$, that is, $t_x t_y = t_{x+y}$. The product $\alpha \cdot x$ of a scalar and a vector gives a vector which corresponds to $t_{\alpha x}$. Now the dilatation $\bar{\alpha}$ and the translation t_x can also be combined by composition; but the ordinary composite $\bar{\alpha}t_x$ is *not* a translation in general since it leaves fixed the vector $(1 - \alpha)^{-1}\alpha x$. However the composite map $\bar{\alpha}t_x\bar{\alpha}^{-1}$ is a translation; in fact $\bar{\alpha}t_x\bar{\alpha}^{-1}(O) = \alpha x$. In other words, the appropriate operation combining a dilatation $\bar{\alpha}$ and a translation t_x is

$$\bar{\alpha} * t_x = \bar{\alpha}t_x\bar{\alpha}^{-1} = t_{\alpha x}.$$

Finally, the two scalars α and β can be added to give $\alpha + \beta$. The corresponding dilatations $\bar{\alpha}$ and $\bar{\beta}$ can be "added" by defining $\bar{\alpha} + \bar{\beta}$ to be that dilatation which maps each vector x to $(\alpha + \beta)x$. Note, however, that this operation differs from the three previous definitions since it is not given entirely within the framework of maps. In terms of maps we could say: Given $\bar{\alpha}$ and $\bar{\beta}$ in $\bar{D}$, let $t_x \in T$. Define $\bar{\alpha} + \bar{\beta}$ to be the dilatation such that

$(\bar{\alpha} + \bar{\beta}) * t_x = (\bar{\alpha} * t_x)(\bar{\beta} * t_x)$. Not only is this definition somewhat awkward, but it is by no means obvious that such a dilatation exists or that it is independent of the choice of t_x. Nevertheless, as we shall soon see, these difficulties can be overcome and this awkward definition proves to be the right one.

These introductory remarks have been intended to suggest that the algebraic structure of a vector space and its division ring can be recovered within the framework of certain maps on the space. We return now to the precise construction in the axiomatic setting.

For the remainder of this section, $\mathscr{P}$ will denote a Desarguian projective space of dimension $n \geq 2$, H some fixed hyperplane of $\mathscr{P}$, and P some fixed point not in H. We let D be the set of all homologies with axis H and center P, and we let E be the set of all elations with axis H. For each $Q \in \mathscr{P}/H$, we shall denote by $\tilde{Q}$ the (unique) elation such that $\tilde{Q}(P) = Q$. In particular $\tilde{P}$ is the identity map on $\mathscr{P}$. Thus we can write $E = \{\tilde{Q}: Q \in \mathscr{P},\ Q \notin H\}$. Because we intend E to play the role of a vector space we shall write $\tilde{Q}_1 \oplus \tilde{Q}_2$ to denote the composite map $Q_1 Q_2$ of two elations.

Lemma 1

$(E, \oplus)$ is an abelian group with identity element $\tilde{P}$.

Proof

By Theorem 5, Section 4.5, $(E, \oplus)$ is a group and clearly $\tilde{P}$ is its identity. It remains to show that the operation is commutative, and we treat this in two cases. First suppose $\tilde{Q}_1$ and $\tilde{Q}_2$ are elements of E with different centers, say S_1, S_2. (See Figure 4.14.) We let $R = (Q_1 + S_2) \cap (Q_2 + S_1)$. By axiom

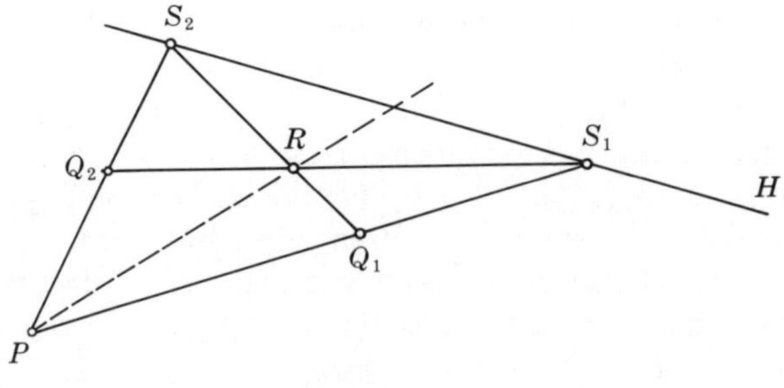

Figure 4.14

PS (2), R is a single point. We claim that

$$\tilde{R} = \tilde{Q}_1 \oplus \tilde{Q}_2 = \tilde{Q}_2 \oplus \tilde{Q}_1.$$

In order to prove this, we need only show that each of the three maps $\tilde{R}$, $\tilde{Q}_1 \oplus \tilde{Q}_2$, and $\tilde{Q}_2 \oplus \tilde{Q}_1$ send the point P to the same image. But by definition, $\tilde{R}(P) = R$. Also $(\tilde{Q}_1 \oplus \tilde{Q}_2)(P) = \tilde{Q}_1(Q_2)$. Now $\tilde{Q}_1(Q_2) \in Q_2 + S_1$

since S_1 is the center of $\tilde{Q}_1$, and $\tilde{Q}_1(Q_2) \in \tilde{Q}_1(P + S_2) = Q_1 + S_2$ since $\tilde{Q}_1$ is a collineation and S_2 is left fixed. Thus $\tilde{Q}_1(Q_2) = R$. Similarly $\tilde{Q}_2(Q_1) = R$ and commutativity holds.

Before treating the case where $\tilde{Q}_1$ and $\tilde{Q}_2$ have the same center, we remark that Figure 4.14 has an intuitive value. It displays the projective analogue of the parallelogram law for addition of vectors: the appropriate "parallelogram" is (P, Q_1, Q_2, R). In terms of $\mathscr{P}/H$, the "sides" $P + Q_1$ and $Q_2 + R$ are parallel since $S_1 \notin \mathscr{P}/H$; likewise the sides $P + Q_2$ and $Q_1 + R$ have no point in common in $\mathscr{P}/H$.

Now suppose $\tilde{Q}_1$ and $\tilde{Q}_2$ are elations with common center S. Figure 4.15

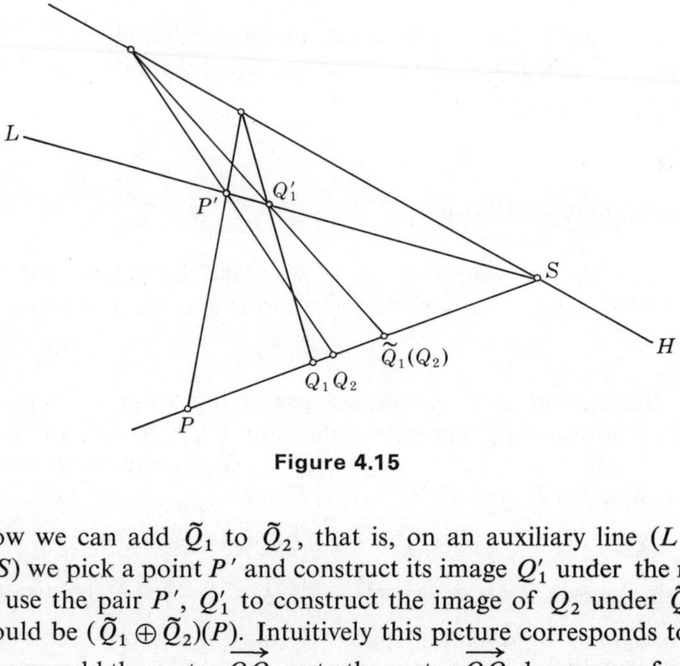

Figure 4.15

shows how we can add $\tilde{Q}_1$ to $\tilde{Q}_2$, that is, on an auxiliary line (L passing through S) we pick a point P' and construct its image Q_1' under the map $\tilde{Q}_1$. Then we use the pair P', Q_1' to construct the image of Q_2 under $\tilde{Q}_1$. That image would be $(\tilde{Q}_1 \oplus \tilde{Q}_2)(P)$. Intuitively this picture corresponds to Figure 4.16 where we add the vector $\overrightarrow{OQ_1}$ onto the vector $\overrightarrow{OQ_2}$ by means of parallelograms.

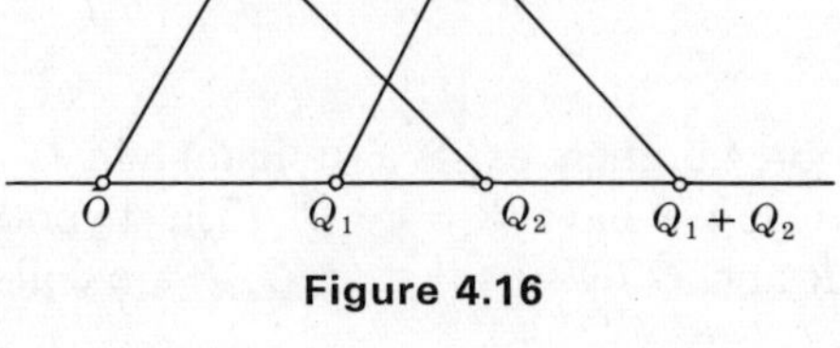

Figure 4.16

Unfortunately it seems that the "proof by picture" technique for this case is more sophisticated than one might expect. However, using the first case we can give a simple algebraic argument. We let $\tilde{Q}_3$ be an elation with center $S' \neq S$ and observe that the composite of two elations with distinct centers has a center distinct from either factor (why?). Thus we have

$$\begin{aligned}\tilde{Q}_3 \oplus (\tilde{Q}_1 \oplus \tilde{Q}_2) &= (\tilde{Q}_3 \oplus \tilde{Q}_1) \oplus \tilde{Q}_2 = \tilde{Q}_2 \oplus (\tilde{Q}_3 \oplus \tilde{Q}_1)\\ &= (\tilde{Q}_2 \oplus \tilde{Q}_3) \oplus \tilde{Q}_1 = (\tilde{Q}_3 \oplus \tilde{Q}_2) \oplus \tilde{Q}_1 = \tilde{Q}_3 \oplus (\tilde{Q}_2 \oplus \tilde{Q}_1).\end{aligned}$$

Since $(E, \oplus)$ is a group, we get

$$\tilde{Q}_1 \oplus \tilde{Q}_2 = \tilde{Q}_2 \oplus \tilde{Q}_1,$$

completing the proof.

The set D will be the nonzero elements of our division ring. To that end we define "multiplication" in D to be the usual composition of functions. From Theorem 5, Section 4.5, we have:

Lemma 2

D is a group under composition.

Before we define "addition" in D, we make the following definition of a product of a scalar times a vector: For $\alpha \in D$ and $\tilde{Q} \in E$ we put

$$\alpha \cdot \tilde{Q} = \alpha \tilde{Q} \alpha^{-1}.$$

Here, as throughout this section, we use juxtaposition of maps solely to indicate composition. By Theorem 5, Section 4.5, $\alpha \cdot \tilde{Q} \in E$. In fact, $\alpha \cdot \tilde{Q}$ is that elation which maps P to $\alpha(Q)$ and $\alpha \cdot \tilde{Q}$ has the same center as $\tilde{Q}$. Moreover, if $\alpha, \beta \in D$, and $\tilde{Q}, \tilde{Q}_1, \tilde{Q}_2 \in E$, then

$$(\alpha\beta) \cdot \tilde{Q} = (\alpha\beta)\tilde{Q}(\alpha\beta)^{-1} = \alpha(\beta\tilde{Q}\beta^{-1})\alpha^{-1} = \alpha \cdot (\beta \cdot \tilde{Q}), \tag{1}$$

$$\alpha \cdot (\tilde{Q}_1 \oplus \tilde{Q}_2) = \alpha(\tilde{Q}_1\tilde{Q}_2)\alpha^{-1} = \alpha\tilde{Q}_1\alpha^{-1}\alpha\tilde{Q}_2\alpha^{-1} = (\alpha \cdot \tilde{Q}_1) \oplus (\alpha \cdot \tilde{Q}_2). \tag{2}$$

Thus two axioms of a vector space relating scalars to vectors are easily established. Important to our development is:

Lemma 3

Let $\tilde{Q}, \tilde{R} \in E$ with $Q \neq P$, $R \neq P$, and suppose S is the common center of $\tilde{Q}$ and $\tilde{R}$. Then there exists a unique $\alpha \in D$ such that $\tilde{R} = \alpha \cdot \tilde{Q}$.

Proof

By Theorem 7, Section 4.5, there exists a (unique) α in D such that $\alpha(Q) = R$. By definition of $\alpha \cdot \tilde{Q}$, we have $\tilde{R} = \alpha \cdot \tilde{Q}$. (The hypothesis that S is the common center of $\tilde{R}$ and $\tilde{Q}$ tells us that P, Q, R are collinear.)

Lemma 3 allows us to define addition on elements of D. Suppose $\alpha, \beta \in D$. Let $\tilde{Q} \in E$ with $Q \neq P$. Then

$$\alpha \cdot \tilde{Q} \oplus \beta \cdot \tilde{Q} = (\alpha \tilde{Q} \alpha^{-1})(\beta \tilde{Q} \beta^{-1}) = \tilde{R}$$

is a composite of two elations with common center, namely the center of $\tilde{Q}$. Hence $\tilde{R}$ has the same center as $\tilde{Q}$ so that (providing $\tilde{R} \neq \tilde{P}$) there exists a unique $\gamma \in D$ such that $\gamma \cdot \tilde{Q} = \tilde{R}$. We define $\alpha + \beta = \gamma$. Once we show that this definition is well defined we have

$$(\alpha + \beta) \cdot \tilde{Q} = \alpha \cdot \tilde{Q} \oplus \beta \cdot \tilde{Q}, \tag{3}$$

which is another axiom of vector spaces we will need.

Lemma 4

The above definition of $\alpha + \beta$ is independent of the choice of $\tilde{Q}$.

Proof

Let $\tilde{S}$ be a second elation with center different from the center of $\tilde{Q}$. Then, as above, there exist $\mu, \nu \in D$ such that

$$\begin{aligned} (\alpha \tilde{S} \alpha^{-1})(\beta \tilde{S} \beta^{-1}) &= \mu \tilde{S} \mu^{-1}, \\ (\alpha \tilde{Q} \tilde{S} \alpha^{-1})(\beta \tilde{Q} \tilde{S} \beta^{-1}) &= \nu \tilde{Q} \tilde{S} \nu^{-1}. \end{aligned} \tag{\dagger}$$

By inserting $1 = \alpha\alpha^{-1} = \beta\beta^{-1} = \nu\nu^{-1}$ into the latter equation, we get

$$(\alpha \tilde{Q} \alpha^{-1})(\alpha \tilde{S} \alpha^{-1})(\beta \tilde{Q} \beta^{-1})(\beta \tilde{S} \beta^{-1}) = (\nu \tilde{Q} \nu^{-1})(\nu \tilde{S} \nu^{-1}).$$

Now using the commutativity of E, we have

$$(\alpha \tilde{Q} \alpha^{-1})(\beta \tilde{Q} \beta^{-1})(\nu \tilde{Q}^{-1} \nu^{-1}) = (\alpha \tilde{S}^{-1} \alpha^{-1})(\beta \tilde{S}^{-1} \beta^{-1})(\nu \tilde{S} \nu^{-1}). \tag{\dagger\dagger}$$

But by Theorem 5, Section 4.5, the product on the left has the same center as $\tilde{Q}$, or else is the identity map. Likewise the product on the right has the same center as $\tilde{S}$, or else is the identity map. But by choice $\tilde{S}$ and $\tilde{Q}$ have distinct centers. This forces each side to be the identity map; hence

$$(\alpha \tilde{S} \alpha^{-1})(\beta \tilde{S} \beta^{-1}) = \nu \tilde{S} \nu^{-1}.$$

By (†), $\nu = \mu$. Also the left-hand side of (††) together with the definition of γ, gives $\mu = \gamma$, and we have shown that the same γ results from choices of $\tilde{Q}$ and $\tilde{S}$ with distinct centers. Since at least two elations exist with different centers, say $\tilde{Q}$ and $\tilde{S}$, any third elation must have a center different from either $\tilde{Q}$ or $\tilde{S}$ and the lemma is proved.

Intuitively our definition of $\alpha + \beta$ makes sense. In any vector space, the sum $\alpha + \beta$ of two scalars is that (unique) scalar such that for any vector $x \neq O$, $(\alpha + \beta)x = \alpha x + \beta x$. There is still one issue to account for. It corresponds to the situation when $\alpha = -\beta$.

We denote by 0 the map from $\mathscr{P}$ to $\mathscr{P}$ which leaves fixed points of H but sends all other points to P, that is,

$$0(Q) = P, \forall Q \notin H.$$

Properly speaking, 0 is not a homology, since it is not 1-1. Nevertheless, we adjoin 0 to D and denote the union by D'. We define:

$$0\alpha = \alpha 0 = 0, \forall \alpha \in D', \tag{4}$$

$$0 \cdot \tilde{Q} = \tilde{P}, \forall \tilde{Q} \in E. \tag{5}$$

We can extend Lemma 3 to read: Let $\tilde{Q}$, $\tilde{R}$ be elations with common center S, and suppose $\tilde{Q} \neq \tilde{P}$. Then there exists a unique $\alpha \in D'$ such that $\tilde{R} = \alpha \cdot \tilde{Q}$. In the case $\tilde{R} = \tilde{P}$, we, of course, take $\alpha = 0$. Now in computing $\alpha + \beta$, for $\alpha, \beta \in D'$ it may happen that $\alpha \cdot \tilde{Q} \oplus \beta \cdot \tilde{Q} = \tilde{P}$, in which case we put $\alpha + \beta = 0$. It follows that

$$\alpha + 0 = 0 + \alpha = 0, \forall \alpha \in D'. \tag{6}$$

In fact, we have

Lemma 5

$(D', +)$ is an abelian group.

Proof

We make use of the fact that $(E, \oplus)$ is an abelian group. Choose $\tilde{Q} \in E$, $\tilde{Q} \neq \tilde{P}$. For $\alpha, \beta, \gamma \in D'$,

$$\begin{aligned}[(\alpha + \beta) + \gamma] \cdot \tilde{Q} &= (\alpha + \beta) \cdot \tilde{Q} \oplus \gamma \cdot \tilde{Q} \\ &= (\alpha \cdot \tilde{Q} \oplus \beta \cdot \tilde{Q}) \oplus \gamma \cdot \tilde{Q} \\ &= \alpha \cdot \tilde{Q} \oplus (\beta \cdot \tilde{Q} \oplus \gamma \cdot \tilde{Q}) \\ &= [\alpha + (\beta + \gamma)] \cdot \tilde{Q}.\end{aligned}$$

Thus addition in D' is associative. Commutativity follows from the commutativity in E, and it is clear that 0 acts as the identity element. We need only establish the existence of (additive) inverses in D'. Suppose $\alpha \in D'$. If $\alpha = 0$, certainly $-\alpha = 0$ works. Hence assume $\alpha \neq 0$. We want $\beta \in D$ such that $\alpha + \beta = 0$, that is, for $\tilde{Q} \in E$, $\alpha \tilde{Q} \alpha^{-1} \beta \tilde{Q} \beta^{-1} = \tilde{P}$, or $\beta \tilde{Q} \beta^{-1} = \alpha \tilde{Q}^{-1} \alpha^{-1}$. Since $\alpha \tilde{Q}^{-1} \alpha^{-1} \in E$ and has the same center as $\tilde{Q}$, the existence (and uniqueness) of such a β is guaranteed by Lemma 3.

Theorem 1

$(D', +, \cdot)$ is a division ring.

Proof

Each of the axioms has either been verified or is quite obvious except for the distributive laws:

$$(\alpha + \beta)\gamma = \alpha\gamma + \beta\gamma, \alpha(\beta + \gamma) = \alpha\beta + \alpha\gamma.$$

We show the first and leave the second to the reader. Pick $\tilde{Q} \neq \tilde{P}$. Then using (1), (3), and (2) we have

$$\begin{aligned}[(\alpha + \beta)\gamma] \cdot \tilde{Q} &= (\alpha + \beta) \cdot (\gamma \cdot \tilde{Q}) \\ &= \alpha \cdot (\gamma \cdot \tilde{Q}) \oplus \beta \cdot (\gamma \cdot \tilde{Q}) \\ &= (\alpha\gamma) \cdot \tilde{Q} \oplus (\beta\gamma) \cdot \tilde{Q} \\ &= (\alpha\gamma + \beta\gamma) \cdot \tilde{Q}.\end{aligned}$$

Hence

$$(\alpha + \beta)\gamma = \alpha\gamma + \beta\gamma.$$

Theorem 2

E is a vector space over D'.

Proof

Theorem 1, Lemma 1, and properties (1) and (2) verify all but the *unitary law* for scalar multiplication: Let α be the identity map and $\tilde{Q} \in E$. Then certainly $\alpha \cdot \tilde{Q} = \alpha\tilde{Q}\alpha^{-1} = \tilde{Q}$, completing the proof.

Theorem 2 does not, as it stands, justify all the work we have done; however it grows in stature as we record:

Theorem 3

The map $Q \to \tilde{Q}$ from $\mathscr{P}/H$ to E is 1-1 and onto and sends each line of $\mathscr{P}/H$ onto an affine line in ${}_{D'}E$, that is, it preserves the affine structure.

Proof

By definition the map is 1-1 and onto. Let Q, S be arbitrary distinct points in $\mathscr{P}/H$. We wish to show that the line $Q + S$ of $\mathscr{P}/H$ is mapped onto the affine line $J(\tilde{Q}, \tilde{S}) = \{\alpha \cdot \tilde{Q} \oplus (1 - \alpha) \cdot \tilde{S} : \alpha \in D'\}$ in ${}_{D'}E$. Without loss of generality, $Q \neq P$.

Case 1

$P \in Q + S$ (that is, $Q + S = P + Q$). Then each $R \in Q + S$ is mapped onto $\tilde{R}$, which has the same center as $\tilde{Q}$ (namely $(P + Q) \cap H$). By Lemma 3, $\tilde{R} = \alpha \cdot \tilde{Q}$ for some $\alpha \in D'$. Moreover for each $\alpha \in D'$, the elation $\alpha \cdot \tilde{Q} = \tilde{R}$

for some $R \in P + Q$. Hence $Q + S = P + Q$ is mapped onto $\{\alpha \cdot \tilde{Q} : \alpha \in D'\}$, which is the affine line $J(\tilde{Q}, \tilde{S})$. (Recall that $\tilde{P}$ plays the role of the identity in ${}_{D'}E$.)

Case 2

$P \notin Q + S$. We put $P_1 = (P + Q) \cap H$, $P_2 = (P + S) \cap H$, and $P_3 = (Q + S) \cap H$ (see Figure 4.17). Then $(P + S) \cap (P_1 + R) = \beta(S)$ for some $\beta \in D'$, and $(P + Q) \cap (P_2 + R) = \alpha(Q)$ for some $\alpha \in D'$. The sum $\beta \cdot \tilde{S} \oplus \alpha \cdot \tilde{Q}$ equals $\tilde{R}$ by definition of addition in E (see Figure 4.14). Also $\alpha(S) = (P + S) \cap$

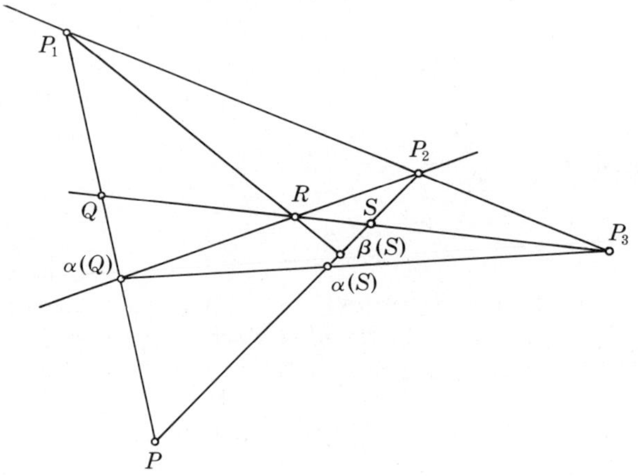

Figure 4.17

$(\alpha(Q) + P_2)$ since α is a homology. Now we add the elations $\alpha \cdot \tilde{S}$ and $\beta \cdot \tilde{S}$ using the auxiliary line $Q + S$ (see Figure 4.15) and observe that $\alpha \cdot \tilde{S} \oplus \beta \cdot \tilde{S} = \tilde{S}$. In other words $(\alpha + \beta) \cdot \tilde{S} = \tilde{S}$, or $\alpha + \beta = 1$, the identity of D'. Thus the point R on $Q + S$ is mapped to $\tilde{R} = \alpha \cdot \tilde{Q} \oplus \beta \cdot \tilde{S}$, where $\alpha + \beta = 1$. To show that the image of $Q + S$ is *all* of $J(\tilde{Q}, \tilde{S})$ essentially amounts to reversing the construction in Figure 4.17, and we leave this as an exercise (see Exercise 4).

We conclude this section with two useful corollaries to Theorem 3.

Corollary 1

Let φ denote the map $Q \to \tilde{Q}$ from $\mathscr{P}/H$ onto ${}_{D'}E$. If A is any flat in $\mathscr{P}/H$, then $\varphi(A)$ is an affine subspace of ${}_{D'}E$. If B is any affine subspace of ${}_{D'}E$, then $\varphi^{-1}(B)$ is a flat in $\mathscr{P}/H$.

Proof

Let A be a flat in $\mathscr{P}/H$ and let $\tilde{Q}_1, \tilde{Q}_2 \in \varphi(A)$ and $\alpha \in D'$. Then $\alpha \cdot \tilde{Q}_1 \oplus (1 - \alpha) \cdot \tilde{Q}_2 = \varphi(R)$ where $R \in Q_1 + Q_2 \subseteq A$ as in the proof of Theorem 3. Similarly, if B is an affine subspace of ${}_{D'}E$, and if $Q_1, Q_2 \in \varphi^{-1}(B)$, then for any $R \in Q_1 + Q_2$ we have

$$\varphi(R) = \alpha \cdot \tilde{Q}_1 \oplus (1 - \alpha) \cdot \tilde{Q}_2 \in B.$$

Corollary 2

$\dim ({}_{D'}E) = p\text{-}\dim (\mathscr{P})$.

Proof

Let $\mathscr{P} = P + P_1 + \cdots + P_n$ where $H = P_1 + \cdots + P_n$, for independent points $\{P, P_1, P_2, \cdots, P_n\}$. For $1 \leq i \leq n$, choose $Q_i \in P + P_i$ such that $Q_i \neq P$, $Q_i \neq P_i$. Then $P \subsetneqq P + Q_1 \subsetneqq P + Q_1 + Q_2 \subsetneqq \cdots \subsetneqq P + Q_1 + \cdots + Q_n = \mathscr{P}/H$ (where these operations are taking place in $\mathscr{P}/H$) implies

$$\varphi(P) \subsetneqq \varphi(P + Q_1) \subsetneqq \varphi(P + Q_1 + Q_2) \subsetneqq \cdots \subsetneqq \varphi(P + Q_1 + \cdots + Q_n) = E.$$

The latter is a chain of affine subspaces of E each containing the zero subspace $\varphi(P)$; hence they are linear subspaces, so that $\dim (E) \geq n$. On the other hand if $\dim (E) > n$, then for at least one k there would exist a linear (and hence affine) subspace B such that

$$\varphi(P + Q_1 + \cdots + Q_k) \subsetneqq B \subsetneqq \varphi(P + Q_1 + \cdots + Q_{k+1}).$$

But, by Corollary 1, $\varphi^{-1}(B)$ is a flat, and we would have

$$P + Q_1 + \cdots + Q_k \subsetneqq \varphi^{-1}(B) \subsetneqq P + Q_1 + \cdots + Q_{k+1},$$

forcing $\varphi^{-1}(B)$ to have dimension strictly greater than k but strictly less than $k + 1$. Hence $\dim (E) = n$.

EXERCISES

1. Let H_1, H_2 be hyperplanes of an n-dimensional Desarguian projective space $\mathscr{P} (n \geq 2)$. Let $P_1 \notin H_1$ and $P_2 \notin H_2$ for points P_1, P_2 of $\mathscr{P}$. Show that there exists a collineation $\psi : \mathscr{P} \to \mathscr{P}$ such that $\psi(H_1) = H_2$ and $\psi(P_1) = P_2$. [*Hint*: Find a common complement H of P_1 and P_2 such that $H \supseteq H_1 \cap H_2$ and construct a perspectivity with axis H which sends P_1 to P_2.]

2. Let $\mathscr{P}$, H_1, H_2, P_1, P_2 be as in Exercise 1. Let ${}_{D_1'}E_1$, ${}_{D_2'}E_2$ be the vector spaces resulting from the construction given in this section using respectively H_1 and P_1 then H_2 and P_2. Show that there exists a collineation (see Exercise 11, Chapter 2, Section 2.6) from ${}_{D_1'}E_1$ onto ${}_{D_2'}E_2$.

3. Let $\mathscr{P} = \mathscr{P}(\mathbb{Z}_3^3)$, and let $H = \{\langle(\alpha, \beta, 0)\rangle : \alpha, \beta \in \mathbb{Z}_3\}$ and $P = \langle(0, 0, 1)\rangle$. Make a list of E and D' in the above context, and show explicitly the collineation from $\mathscr{P}/H$ onto $\mathbb{Z}_3^2$.

4. Complete the proof of Theorem 3. [*Hint*: Define P_1, P_2, P_3 as in Figure 4.17. Now construct $\beta(S)$, $\alpha(S)$ and check that if $\beta \cdot \tilde{S} \oplus \alpha \cdot \tilde{S} = \tilde{S}$, then $\beta(S) + P_1$ must meet $\alpha(Q) + P_2$ on $Q + S$.]

5. A more sophisticated approach to showing that D' (as defined in this section) is a division ring can be pursued as follows. View D' as a subset of Hom (E, E), the ring of homomorphisms of the abelian group E into itself, that is $\alpha(\tilde{Q}) = \alpha\tilde{Q}\alpha^{-1}$, for $\alpha \in D'$, $\tilde{Q} \in E$. Then prove that D' is a *subring* by showing closure under multiplication (trivial) and addition. Note that addition of homomorphisms is always *defined* as follows: If $\alpha, \beta \in \text{Hom}(E, E)$ and $\tilde{Q} \in E$, then $(\alpha + \beta)(\tilde{Q}) = \alpha(\tilde{Q}) \oplus \beta(\tilde{Q})$. This approach to the problem eliminates the verification of many of the division ring axioms, since subrings "inherit" associativity and distributivity automatically.

SECTION 4.7

COORDINATIZATION OF $\mathscr{P}$ AND THE FUNDAMENTAL THEOREM OF PROJECTIVE GEOMETRY

In this section we shall make use of the vector space construction relative to $\mathscr{P}/H$ to show that $\mathscr{P} \approx \mathscr{P}(D^{n+1})$ for some division ring D. This will enable us to characterize the collineations from $\mathscr{P}$ onto $\mathscr{P}'$ in terms of collineations on their underlying vector spaces. Our first step is to show how an affine space can be imbedded into a projective space. This process makes precise the aforementioned construction by adjoining "points at infinity."

Let D be a division ring and let $V = D^n$ for some $n \geq 2$. Let $\mathscr{P} = \mathscr{P}(D^{n+1})$. To each $x \in V$, we associate a point $\varphi(x)$ in $\mathscr{P}$ as follows:

$$\varphi((\alpha_1, \cdots, \alpha_n)) = \langle(\alpha_1, \cdots, \alpha_n, 1)\rangle.$$

This map φ is clearly 1-1 since

$$\langle(\alpha_1, \cdots, \alpha_n, 1)\rangle = \langle(\beta_1, \cdots, \beta_n, 1)\rangle \Rightarrow (\alpha_1, \cdots, \alpha_n, 1) = \gamma(\beta_1, \cdots, \beta_n, 1)$$

for some $\gamma \in D \Rightarrow \gamma = 1 \Rightarrow \alpha_i = \beta_i$, $i = 1, 2, \cdots, n$. If we let

$$H = \{\langle(\gamma_1, \gamma_2, \cdots, \gamma_n, 0)\rangle : \gamma_i \in D\},$$

then H is a hyperplane of $\mathscr{P}$ since $H \neq \mathscr{P}$, and H contains the n independent points $\langle(1, 0, \cdots, 0)\rangle$, $\langle(0, 1, 0, \cdots, 0)\rangle, \cdots, \langle(0, 0, \cdots, 0, 1, 0)\rangle$. Moreover any point of $\mathscr{P}$ not in H must be of the form $P = \langle(\alpha_1, \cdots, \alpha_n, \alpha)\rangle$ where $\alpha \neq 0$, so that

$$P = \langle(\alpha^{-1}\alpha_1, \cdots, \alpha^{-1}\alpha_n, 1)\rangle = \varphi((\alpha^{-1}\alpha_1, \cdots, \alpha^{-1}\alpha_n)).$$

Thus φ maps V onto $\mathscr{P}/H$, the associated affine space of $\mathscr{P}$ relative to H. Let $x = (\alpha_1, \cdots, \alpha_n)$, $y = (\beta_1, \cdots, \beta_n)$, and let $\alpha \in D$: $\alpha x + (1 - \alpha)y = (\alpha\alpha_1 + (1 - \alpha)\beta_1, \cdots, \alpha\alpha_n + (1 - \alpha)\beta_n)$. Then

$$\begin{aligned}
\varphi(\alpha x + (1-\alpha)y) &= \langle(\alpha\alpha_1 + (1-\alpha)\beta_1, \cdots, \alpha\alpha_n + (1-\alpha)\beta_n, \alpha + (1-\alpha)1)\rangle \\
&= \langle\alpha(\alpha_1, \cdots, \alpha_n, 1) + (1-\alpha)(\beta_1, \cdots, \beta_n, 1)\rangle \\
&\in \langle\alpha(\alpha_1, \cdots, \alpha_n, 1)\rangle + \langle(1-\alpha)(\beta_1, \cdots, \beta_n, 1)\rangle \\
&= \langle(\alpha_1, \cdots, \alpha_n, 1)\rangle + \langle(\beta_1, \cdots, \beta_n, 1)\rangle \\
&= \varphi(x) + \varphi(y).
\end{aligned}$$

On the other hand if $Q \in \varphi(x) + \varphi(y)$, then

$$Q = \langle\alpha(\alpha_1, \cdots, \alpha_n, 1) + \beta(\beta_1, \cdots, \beta_n, 1)\rangle$$

and, if we suppose $Q \notin H$, then $\gamma = \alpha + \beta \neq 0$. Hence, letting $\alpha' = \gamma^{-1}\alpha$ and $\beta' = \gamma^{-1}\beta$ we have $\alpha' + \beta' = 1$ and we can write

$$\begin{aligned}
Q &= \langle\alpha'(\alpha_1, \cdots, \alpha_n, 1) + (1-\alpha')(\beta_1, \cdots, \beta_n, 1)\rangle \\
&= \langle(\alpha'\alpha_1 + (1-\alpha')\beta_1, \cdots, \alpha'\alpha_n + (1-\alpha')\beta_n, 1)\rangle \\
&= \varphi(\alpha' x + (1-\alpha')y).
\end{aligned}$$

Thus φ sends each affine line of V onto a line of $\mathscr{P}/H$. If we think of V as being identified with its image $\varphi(V)$ in $\mathscr{P}$, then we can call the points of H "points at infinity." To each line L of V we adjoin *one* point at infinity, namely the unique point Q of H such that $\varphi(L) \cup \{Q\}$ is a line in $\mathscr{P}$. Now it is the structure of $\mathscr{P}$ which guarantees that points at infinity behave precisely like other points in $\mathscr{P}$ with regard to incidence relations.

Figure 4.18 illustrates what this process looks like when $V = \mathbb{R}^2$ and $\mathscr{P} = \mathscr{P}(\mathbb{R}^3)$. Intuitively we think of placing a copy of $\mathbb{R}^2$ on the plane $A = \{(\alpha, \beta, 1) : \alpha, \beta \in \mathbb{R}\}$ and associating with each point in this plane that unique line through (0, 0, 0) which passes through the point. The xy-plane represents H, the hyperplane of points at infinity. If, for example, two lines L

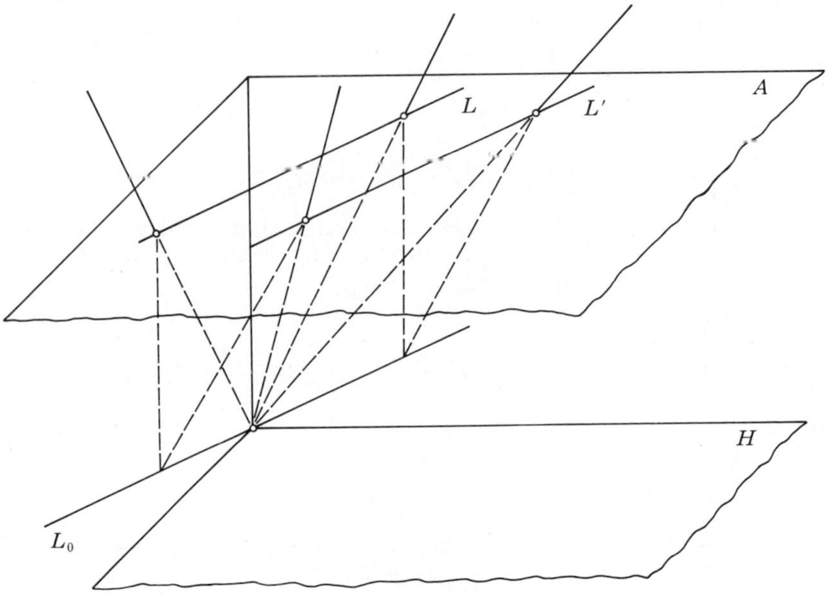

Figure 4.18

and L' of A are parallel, then the associated lines in $\mathscr{P}(\mathbb{R}^3)$ will be two affine planes which meet in an affine line L_0 in H, that is, the lines L and L' meet in the point at infinity L_0.

Now suppose $\mathscr{P}$ is a projective space of dimension $n \geq 2$, and that H, P, E, D' are respectively the hyperplane, point, vector space, and division ring as discussed in the last section. Then dim $(E) = n$, and if $\{x_1, \cdots, x_n\}$ is some basis of E, we can define a map $\varphi: V \to \mathscr{P}(D^{n+1}) = \mathscr{P}'$ by

$$\varphi\left(\sum_{i=1}^{n} \alpha_i x_i\right) = \langle(\alpha_1, \alpha_2, \cdots, \alpha_n, 1)\rangle.$$

As above, the map φ is 1-1 from V onto $\mathscr{P}'/H'$, where

$$H' = \{\langle(\alpha_1, \cdots, \alpha_n, 0)\rangle : \alpha_1 \in D\}$$

and φ maps each line A of V onto a line L'/H' of $\mathscr{P}'/H'$ (where L' is a line in $\mathscr{P}'$). If we let $\psi: \mathscr{P}/H \to E$ be the map which sends Q to $\tilde{Q}$ (as defined in the last section), then the composite $\tau = \varphi \cdot \psi: \mathscr{P}/H \to \mathscr{P}'/H'$ is 1-1 and onto and sends each line L/H of $\mathscr{P}/H$ onto a line L'/H' of $\mathscr{P}'/H'$ (since each factor of the composite has these properties). In short, τ is a collineation from $\mathscr{P}/H$ onto $\mathscr{P}'/H'$. We wish to extend τ to a collineation $\tau': \mathscr{P} \to \mathscr{P}'$.

Let $\tau(P) = P'$. For each $Q \in H$, there exists a unique line L of $\mathscr{P}$ containing P and Q, namely $L = P + Q$. Then $\tau(L/H) = L'/H'$ for some line L' of $\mathscr{P}'$. We let $Q' = L' \cap H'$. Since L'/H' is a bonafide *line* of $\mathscr{P}'/H'$ (that is, $L' \nsubseteq H'$), we see that Q' is a point of H'. We define $\tau'(Q) = Q'$. For $R \in \mathscr{P}/H$, we define $\tau'(R) = \tau(R)$. Then τ' is defined on all of $\mathscr{P}$.

Theorem 1

The map $\tau': \mathscr{P} \to \mathscr{P}'$ is a collineation.

Proof

We first show that τ' is 1-1 and onto. Let $Q_1, Q_2 \in H$ with $Q_1 \neq Q_2$. Then $L_1 = P + Q_1$ and $L_2 = P + Q_2$ are distinct lines in $\mathscr{P}$. Hence $\tau(L_1/H) = L_1'/H'$ and $\tau(L_2/H) = L_2'/H'$ are distinct lines of $\mathscr{P}'/H'$, both containing P'. Thus the points $\tau'(Q_1) = L_1' \cap H$ and $\tau'(Q_2) = L_2' \cap H$ must be distinct points. If $Q_1 \in H$ and $Q_2 \notin H$, then $\tau'(Q_1) \in H'$ and $\tau'(Q_2) \notin H'$ so that $\tau'(Q_1) \neq \tau'(Q_2)$. Finally if $Q_1 \notin H$ and $Q_2 \notin H$ with $Q_1 \neq Q_2$, then $\tau'(Q_1) = \tau(Q_1) \neq \tau(Q_2) = \tau'(Q_2)$. These cases show that τ' is 1-1. To show that τ' is onto, it suffices to show that each point $Q' \in H'$ is the image of some point (in H). Suppose $Q' \in H'$. Let $L' = P' + Q'$. Then $L'/H' = \tau(L/H)$ for some line L of $\mathscr{P}$ since τ is onto and maps lines onto lines. Letting $Q = L \cap H$, we see that $\tau'(Q) = Q'$, and τ' is onto.

Finally we must show that τ' maps each line of $\mathscr{P}$ onto a line of $\mathscr{P}'$. From the definition of τ' if L is a line of $\mathscr{P}$ such that $L \cap H$ is a point, then $\tau'(L)$ is a line of $\mathscr{P}'$ (such that $\tau'(L) \cap H'$ is a point). It remains to show that if $L \subseteq H$ for some line L of $\mathscr{P}$, then $\tau'(L)$ is a line of $\mathscr{P}'$. Let L be such a line, and let Q_1, Q_2, Q_3 be distinct points of L (see Figure 4.19).

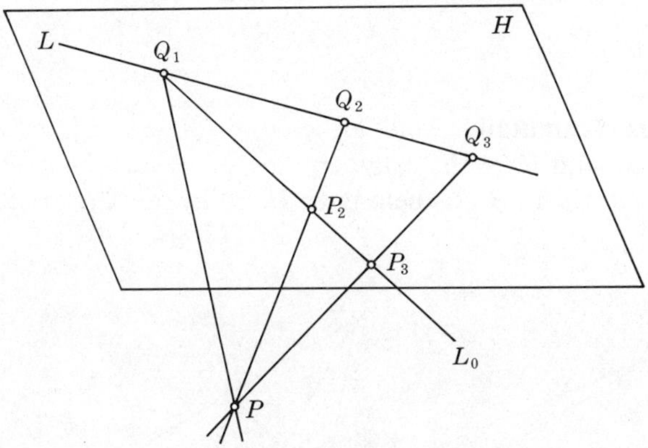

Figure 4.19

Let $P_2 \in P + Q_2$, $P_2 \neq P$, $P_2 \neq Q_2$. Then the line $L_0 = Q_1 + P_2$ meets $P + Q_3$ in some point P_3, $P_3 \neq Q_3$, $P_3 \neq P$. The three image lines $\tau'(P + Q_1)$, $\tau'(P + Q_2)$, and $\tau'(P + Q_3)$ must then be coplanar since each one passes through $P' = \tau'(P)$ and each one intersects $\tau(L_0)$. Hence the plane these image lines determine meets H' in a line L', which by definition of τ' contains $\tau'(Q_1)$, $\tau'(Q_2)$, and $\tau'(Q_3)$. We conclude $\tau'(L) \subseteq L'$. To show equality, let $Q' \in L'$. Then $P' + Q' = \tau(P + Q)$ for some $Q \in H$. But $Q' \in L' \Rightarrow P' + Q'$ meets $\tau'(L_0) \Rightarrow P + Q$ meets $L_0 \Rightarrow Q \in L$, and the proof is complete.

Thus each finite dimensional Desarguian space $\mathscr{P}$ is projectively equivalent to $\mathscr{P}(D^{n+1})$ for some division ring D. Before we exploit this fact to investigate collineations, we pause to record an interesting result that further illustrates the way a geometric property of $\mathscr{P}$ can influence the algebraic structure of a model $\mathscr{P}(D^{n+1})$ of $\mathscr{P}$.

DEFINITION A space (affine or projective) is called *Pappian* (after Pappus of Alexandria) if the following condition holds.

Let L, L' be distinct coplanar lines and suppose A, B, C are distinct points on L not on L' and A', B', C' are distinct points on L' not on L. Then the three points $D = (A + B') \cap (A' + B)$, $E = (A + C') \cap (A' + C)$, and $F = (B + C') \cap (B' + C)$ are collinear. (In the affine case, we *assume* that D, E, F are points.)

The following result partially justifies our reluctance to work over a field rather than a division ring by showing that commutativity of multiplication presupposes a stronger geometric condition than Desargues' theorem.

Theorem 2

Let $\mathscr{P}$ be a Desarguian space of dimension $n \geq 2$. Then $\mathscr{P}$ is Pappian if and only if the associated division ring D' is a field.

Proof

Suppose D' is commutative, and let A, B, C, A', B', C', D, E, F be given. Put $P = L \cap L'$ and let H be a hyperplane of $\mathscr{P}$ such that $D + F \subseteq H$, but none of P, A, B, C, A', B', C' belong to H. (Why must H exist?) Let α and β

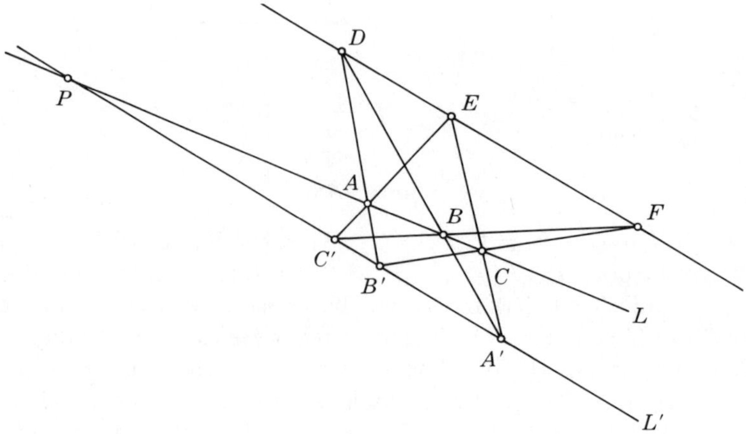

Figure 4.20

be the homologies defined by $\alpha(A) = B$ and $\beta(B) = C$. Then (consult Figure 4.20) $\alpha(B') = A'$ and $\beta(C') = B'$. Hence $\alpha\beta(C') = A'$ and $\beta\alpha(A) = C$. Since $\alpha\beta = \beta\alpha$, the homology sending C' to A' must send A to C. But this forces $(A + C) \cap (A' + C') = E$ to lie in H and hence on $D + F$ since all the points involved are coplanar.

Conversely suppose $\mathscr{P}$ is Pappian. Let α, β be homologies, and let L, L' be distinct lines through P, the center of α and β. Let H be the axis of α and β and choose $A \in L$, $A \notin H$, $A \neq P$, and put $B = \alpha(A)$ and $C = \beta(B)$. On L' let C' be distinct from P and not in H. Put $B' = \beta(C')$ and $A' = \alpha(B')$. Letting D, E, F be defined as in the definition of Pappian space, we have that D, E, and F are collinear. By our construction $D \in H$ and $F \in H$ (see Figure 4.20) and hence $E \in H$. But then $\beta\alpha$, which sends A to C, leaves E fixed and so sends C' to A'; that is, $\beta\alpha = \alpha\beta$, and commutativity of D' is established.

The remainder of this section is devoted to the investigation of collineations from $\mathscr{P}(V)$ to $\mathscr{P}(V')$, where V and V' are vector spaces over division rings D and D', respectively. So far our examples have consisted of perspectivities, in case $V = V'$, and of maps induced from linear, or semilinear, maps from V onto V'. For instance we define $f: {}_{\mathbb{C}}\mathbb{C}^3 \to {}_{\mathbb{C}}\mathbb{C}^3$ by $(\alpha_1, \alpha_2, \alpha_3)f = (\bar{\alpha}_1, \bar{\alpha}_2, \bar{\alpha}_3)$, where $\bar{\alpha}$ denotes the complex conjugate of α. Then the induced map $\hat{f}: \mathscr{P}(\mathbb{C}^3) \to \mathscr{P}(\mathbb{C}^3)$ given by

$$\hat{f}(\langle(\alpha_1, \alpha_2, \alpha_3)\rangle) = \langle(\bar{\alpha}_1, \bar{\alpha}_2, \bar{\alpha}_3)\rangle$$

is a collineation since f is semilinear. Moreover it can be shown (Exercise 1) that f is not induced from any linear map $g: \mathbb{C}^3 \to \mathbb{C}^3$. We will prove, however, that each collineation is induced by a semilinear map on the underlying vector spaces.

Henceforth when we work with a projective space $\mathscr{P}(V)$, we make the convention that we write $\langle x \rangle$ for $x \in V$ only when $x \neq O$. In this case $\langle x \rangle$ is a bona fide point of $\mathscr{P}(V)$, and we sometimes refer to x as a *representative* of the point $\langle x \rangle$.

Suppose $\varphi: \mathscr{P}(V) \to \mathscr{P}(V')$ is a collineation and $\dim V (= \dim V') = n \geq 3$. We establish some lemmas in this context.

Lemma 1

Given x, $y \in V$, and $\langle x \rangle \neq \langle y \rangle$. Let x' be such that $\varphi(\langle x \rangle) = \langle x' \rangle$. Then there exists a unique $y' \in V'$ such that

$$\varphi(\langle y \rangle) = \langle y' \rangle \text{ and } \varphi(\langle x - y \rangle) = \langle x' - y' \rangle.$$

Proof

Let $y_1 \in \varphi(\langle y \rangle)$, that is, $\langle y_1 \rangle = \varphi(\langle y \rangle)$. Then

$$\varphi(\langle x - y \rangle) \subseteq \varphi(\langle x \rangle + \langle y \rangle) = \langle x' \rangle + \langle y_1 \rangle \Rightarrow \varphi(\langle x - y \rangle) = \langle \alpha x' + \beta y_1 \rangle.$$

But since $\langle x-y\rangle \neq \langle x\rangle$ and $\langle x-y\rangle \neq \langle y\rangle$, we have $\alpha \neq 0$, $\beta \neq 0$. Thus

$$\varphi(\langle x-y\rangle) = \langle \alpha x' + \beta y_1\rangle = \langle x' + \alpha^{-1}\beta y_1\rangle.$$

Putting $y' = -\alpha^{-1}\beta y_1$, we have the desired result.

Whenever x, x', y satisfy the hypothesis of Lemma 1 we shall denote y' by $y' = h(x, x', y)$.

Lemma 2

Suppose x, y, z are linearly independent vectors in V, and $x' \in \varphi(\langle x\rangle)$, $y' = h(x, x', y)$. Then $h(x, x', z) = h(y, y', z)$.

Proof

Put $z' = h(x, x', z)$. Then

$$\varphi(\langle y-z\rangle) \subseteq \varphi(\langle y\rangle) + \varphi(\langle z\rangle) = \langle y'\rangle + \langle z'\rangle.$$

Hence, $\varphi(\langle y-z\rangle) = \langle \alpha y' + \beta z'\rangle$ for some $\alpha, \beta \in D'$. Also,

$$\begin{aligned}\varphi(\langle y-z\rangle) = \varphi(\langle y-x+x-z\rangle) &\subseteq \varphi(\langle y-x\rangle) + \varphi(\langle x-z\rangle)\\ &= \langle x'-y'\rangle + \langle x'-z'\rangle.\end{aligned}$$

Hence $\varphi(\langle y-z\rangle) = \langle \gamma(x'-y') + \delta(x'-z')\rangle$ for some $\gamma, \delta \in D'$. Then for some $\mu \in D'$,

$$\alpha y' + \beta z' = \mu(\gamma(x'-y') + \delta(x'-z')).$$

Since $\{x', y', z'\}$ are linearly independent (why?), we have $\gamma + \delta = 0$, and $\alpha y' + \beta z' = -\mu \cdot \gamma \cdot (y'-z')$, giving

$$\varphi(\langle y-z\rangle) = \langle y'-z'\rangle, \text{ that is, } h(y, y', z) = z'.$$

Theorem 3 (Fundamental Theorem of Projective Geometry)

Let $\varphi: \mathscr{P}(V) \to \mathscr{P}(V')$ be a collineation where V, V' are vector spaces over division rings D, D' respectively, with $\dim V = n \geq 3$. Then there exist a semilinear map $f: V \to V'$ such that $\varphi = \hat{f}$, that is, $\varphi(\langle x\rangle) = \langle (x)f\rangle$, $\forall x \in V$. If in addition $\varphi = \hat{g}$, for another semilinear map $g: V \to V'$, then for each $x \in V$, $(x)g = \alpha(x)f$ for some fixed $\alpha \in D'$, and the division ring isomorphism associated with g is τ where $(\beta)^\tau = (\alpha\beta\alpha^{-1})^\sigma$ and σ is the isomorphism associated with f.

Proof

Let $\{x, y, z\}$ be linearly independent vectors in V, and let x', y', z' be as in Lemma 2. For $w \in V$, we define $(w)f$ as follows: at most one of the pairs $\{x, w\}$, $\{y, w\}$, $\{z, w\}$ can be linearly dependent. If, say, $\{x, w\}$ is linearly independent, we define $(w)f = h(x, x', w)$.

Claim 1

f is well defined. Suppose $\{x, w\}$ and $\{y, w\}$ are both linearly independent. Then we want $h(x, x', w) = h(y, y', w)$. There are two cases to consider. Either $\{x, y, w\}$ is linearly independent, and Lemma 2 yields $h(x, x', w) = h(y, y', w)$; or $\{x, y, w\}$ is linearly dependent in which case $w \in \langle x, y\rangle$ and hence $w \notin \langle y, z\rangle$ and $w \notin \langle x, z\rangle$, that is, $\{y, z, w\}$ and $\{x, z, w\}$ are linearly independent. Then by Lemma 2, $h(y, y', w) = h(z, z', w)$ and $h(z, z', w) = h(x, x', w)$, so that $h(x, x', w) = h(y, y', w)$.

Claim 2

f is 1-1. Suppose $w' = h(x, x', w_1) = h(x, x', w_2)$. Then in particular $\varphi(\langle w_1\rangle) = \langle w'\rangle = \varphi(\langle w_2\rangle)$. Since φ is 1-1, $w_2 = \alpha w_1$, for some $\alpha \in D'$. Also,

$$\varphi(\langle x - w_1\rangle) = \langle x' - w'\rangle = \varphi(\langle x - \alpha w_1\rangle) \Rightarrow \langle x - w_1\rangle = \langle x - \alpha w_1\rangle \Rightarrow x - w_1 = \beta(x - \alpha w_1).$$

But $\{x, w_1\}$ is linearly independent so that $\beta = 1$ and then $\alpha = 1$, that is, $w_2 = w_1$ and f is 1-1.

Claim 3

f is onto. Let $u' \in V'$. There exists $u_1 \in V$ such that $\varphi(\langle u_1\rangle) = \langle u'\rangle$. Without loss of generality, $\{x, u_1\}$ is linearly independent. Hence

$$\varphi(\langle x\rangle + \langle u_1\rangle) = \langle x'\rangle + \langle u'\rangle \Rightarrow \exists \alpha, \beta \in D$$

such that $\varphi\langle \alpha x + \beta u_1\rangle = \langle x' - u'\rangle$. Putting $u = -\alpha^{-1}\beta u_1$, we see that $(u)f = u'$.

Claim 4

f sends each line of V onto a line of V'. Let L be a line of V. Clearly if $O \in L$, then $L = \langle w\rangle$ for some $w \in V$ and $\alpha w \in L \Rightarrow (\alpha w)f \in \varphi(\langle w\rangle) = \langle w'\rangle$. Using the above argument that f is onto, we see that $(L)f = \langle w'\rangle = L'$, a line in V'. Now suppose that L is a line in V not containing O. Let w_1, w_2 be distinct vectors in L, and without loss of generality, $x \notin \langle w_1, w_2\rangle$. Then letting $w_1' = (w_1)f$ and $w_2' = (w_2)f$, we take an arbitrary vector $w_3 = \alpha w_1 + (1 - \alpha) w_2$, $\alpha \neq 0, 1$ on L, and compute $\varphi(\langle x - w_3\rangle) = \langle x' - w_3'\rangle$, where $w_3' = (w_3)f$. But

$$\begin{aligned}\varphi(\langle x - w_3\rangle) &= \varphi(\langle x - \alpha w_1 - (1 - \alpha)w_2\rangle)\\ &= \varphi(\langle \alpha x - \alpha w_1 + (1 - \alpha)x - (1 - \alpha)w_2\rangle)\\ &\subseteq \varphi(\langle \alpha x - \alpha w_1\rangle) + \varphi(\langle (1 - \alpha)x - (1 - \alpha)w_2\rangle)\\ &= \langle x' - w_1'\rangle + \langle x' - w_2'\rangle.\end{aligned}$$

Hence $\langle x' - w_3' \rangle = \langle \gamma(x' - w_1') + \delta(x' - w_2') \rangle$, or $x' - w_3' = \alpha(x' - w_1') + \beta(x' - w_2')$. Since $x' \notin \langle w_1, w_2 \rangle$ (why?), we must have $\alpha + \beta = 1$, giving

$$-w_3' = -\alpha w_1' - (1 - \alpha) w_2'$$

or

$$w_3' = \alpha w_1' + (1 - \alpha) w_2' \in J(w_1', w_2').$$

Thus f sends $J(w_1, w_2)$ into $J(w_1', w_2')$.

To show that f maps $J(w_1, w_2)$ *onto* $J(w_1', w_2')$ we let $w' \in J(w_1', w_2')$. Then $w' = \langle w' \rangle \cap J(w_1', w_2')$. But $w' = (w_0)f$ for some $w_0 \in V$ and

$$\langle w' \rangle \subseteq \langle w_1' \rangle + \langle w_2' \rangle \Rightarrow \langle w_0 \rangle \subseteq \langle w_1 \rangle + \langle w_2 \rangle.$$

Letting $w = \langle w_0 \rangle \cap J(w_1, w_2)$, we have $(w)f \in \varphi(\langle w_0 \rangle) \cap J(w_1', w_2') = w'$, and the result is established.

We have shown that $f: V \to V'$ is a collineation. It follows from Chapter 2, Section 2.6, Exercise 11 that f is semilinear. Obviously $\hat{f} = \varphi$ by definition of f. The remaining assertion of the theorem is left to the reader (Exercise 2).

Now that we know each collineation is induced from a semilinear map, we are naturally led to ask whether each collineation $\varphi: \mathscr{P}(V) \to \mathscr{P}(V)$ is a perspectivity, since perspectivities, to this point, have played such an important role. Also it is reasonable to ask whether we can distinguish those maps φ that are induced by *linear* maps on V from the others. The proofs of the following results are left to the exercises.

Theorem 4

If $\varphi: \mathscr{P}(V) \to \mathscr{P}(V)$ is a collineation having at least one line of fixed points, then $\varphi = \hat{g}$ for some linear map $g: V \to V$.

Theorem 5

If $\varphi: \mathscr{P}(V) \to \mathscr{P}(V)$ with $\varphi = \hat{g}$ for some linear map g, then φ can be written as the product of at most $n + 1$ perspectivities, where $\dim(V) = n + 1$.

Remark: Although the literature is somewhat inconsistent on this point, we shall refer to a collineation induced by a *linear* map as a *projective map* or a *projectivity*. We can restate Theorems 4 and 5 as:

a collineation having a line of fixed points is a projectivity;
each projectivity is a product of perspectivities.

Unlike linear maps, a collineation is not uniquely determined by its action on a basis of the space. For example, the map φ defined earlier by

$$\varphi(\langle(\alpha_1, \alpha_2, \alpha_3)\rangle) = \langle(\bar{\alpha}_1, \bar{\alpha}_2, \bar{\alpha}_3)\rangle \text{ on } \mathscr{P}(\mathbb{C}^3)$$

leaves fixed the points $\langle(1, 0, 0)\rangle$, $\langle(0, 1, 0)\rangle$, $\langle(0, 0, 1)\rangle$, and $\langle(1, 1, 1)\rangle$, any three of which form a basis of $\mathscr{P}(\mathbb{C}^3)$, and yet φ is not the identity map. The following two theorems, however, provide useful analogues to the linear map situation.

Theorem 6

Let $\dim V = n + 1 \geq 3$, and let $\{P_0, P_1, \cdots, P_{n+1}\}$ and $\{Q_0, Q_1, \cdots, Q_{n+1}\}$ be sets of points in $\mathscr{P}(V)$ such that in each set no $n + 1$ of the points lie in a hyperplane. Then there exists a collineation $\varphi:\mathscr{P}(V) \to \mathscr{P}(V)$ such that $\varphi(P_i) = Q_i$, $i = 0, 1, \cdots, n + 1$. In fact φ can be chosen to be a projectivity.

Proof

The hypothesis guarantees that if $P_i = \langle x_i \rangle$ and $Q_i = \langle y_i \rangle$ for $i = 0, 1, \cdots, n + 1$, then $\{x_0, \cdots, x_n\}$ and $\{y_0, \cdots, y_n\}$ are bases of V. Then $x_{n+1} = \sum_{i=0}^{n} \alpha_i x_i$ and $y_{n+1} = \sum_{i=0}^{n} \beta_i y_i$. Moreover none of the α_i's or β_i's are 0 (why?). In other words $\{\alpha_0 x_0, \cdots, \alpha_n x_n\}$ and $\{\beta_0 y_0, \cdots, \beta_n y_n\}$ are also bases. Let $g:V \to V$ be defined by

$$(\alpha_i x_i)g = \beta_i y_i, \quad i = 0, 1, \cdots, n.$$

Then $(x_{n+1})g = y_{n+1}$ and $\hat{g}:\mathscr{P}(V) \to \mathscr{P}(V)$ is the desired map.

Theorem 7

Let D be a commutative field and $\dim ({}_D V) = n + 1$. If g is a linear map on V such that $\hat{g}$ leaves fixed $n + 2$ points of $\mathscr{P}(V)$ no $n + 1$ of which lie in a hyperplane, then $\hat{g}$ is the identity map.

Proof

Let $P_0, P_1, \cdots, P_{n+1}$ be points satisfying the hypothesis, and let $P_i = \langle x_i \rangle$, $i = 0, 1, \cdots, n + 1$. Then $\{x_0, x_1, \cdots, x_n\}$ is a basis of V. Suppose $(x_i)g = \alpha_i x_i$, $i = 0, 1, \cdots, n + 1$, and let $x_{n+1} = \sum_{i=0}^{n} \beta_i x_i$. Then $\alpha_{n+1} x_{n+1} = (x_{n+1})g = \sum_{i=0}^{n} \alpha_i \beta_i x_i$. By uniqueness of representation, $\alpha_{n+1}\beta_i = \alpha_i \beta_i$, $i = 0, 1, \cdots, n$. Hence $\alpha_i = \alpha_{n+1}$, $i = 0, 1, \cdots, n$. But then for arbitrary $x \in V$,

$$x = \sum_{i=0}^{n} \gamma_i x_i \Rightarrow (x)g = \sum_{i=0}^{n} \gamma_i \alpha_{n+1} x_i = \alpha_{n+1} x.$$

Thus $\langle (x)g \rangle = \langle x \rangle$, $\forall x$ and $\hat{g} = 1$.

Corollary

If D is commutative and f, g are linear maps on ${}_DV$ such that $\hat{f}$ and $\hat{g}$ agree on $n+2$ points of $\mathscr{P}(V)$ no $n+1$ of which lie in a hyperplane, then $\hat{f} = \hat{g}$.

We conclude this section with some remarks about coordinates of a projective space. First of all let us give a formal definition. If $\mathscr{P}$ is a projective space (of dimension n) then a *coordinatization* of $\mathscr{P}$ is any collineation $\varphi : \mathscr{P} \to \mathscr{P}(D^{n+1})$, for some division ring D. The coordinatizations that will prove fruitful in the next two sections arise in a special situation. Suppose D is commutative and $\mathscr{P} \approx \mathscr{P}(D^{n+1})$. Let $P_0, P_1, \cdots, P_{n+1}$ be points of $\mathscr{P}$ no $n+1$ of which lie in a hyperplane. By Theorem 6 and Theorem 7 there is a unique projectivity $\varphi : \mathscr{P} \to \mathscr{P}(D^{n+1})$ such that $\varphi(P_i) = \langle e_i \rangle, i = 1, 2, \cdots, n+1$ and $\varphi(P_0) = \langle \sum_{i=1}^{n+1} e_i \rangle$, where $\{e_1, \cdots, e_{n+1}\}$ is the standard basis of D^{n+1}. This set of points $\{P_0, P_1, \cdots, P_{n+1}\}$ is called a *frame of reference* for $\mathscr{P}$.

EXERCISES

1. Show that the map $\varphi : \mathscr{P}(\mathbb{C}^3) \to \mathscr{P}(\mathbb{C}^3)$ given by $\varphi(\langle(\alpha_1, \alpha_2, \alpha_3)\rangle) = \langle(\bar{\alpha}_1, \bar{\alpha}_2, \bar{\alpha}_3)\rangle$ is a collineation that is not induced from any linear map on $\mathbb{C}^3$. Hence, in particular, $\varphi \neq 1$.

2. Show that if $f, g : V \to V'$ are semilinear maps such that $\hat{f} = \hat{g}$ then $\exists \alpha \in D'$ such that $(x)g = \alpha(x)f$, $\forall x \in V$. Now complete the proof of Theorem 3. [*Hint*: consider $\hat{f}(\langle x \rangle)$, $\hat{f}(\langle y \rangle)$, and $\hat{f}(\langle x - y \rangle)$ for independent vectors x, y of V.]

3. Prove Theorem 4.

4. Let $U < V$, $x, y \in V$, $x, y \notin U$. Show that there is a linear map (isomorphism) $f : V \to V$ such that $\hat{f}$ is a perspectivity on $\mathscr{P}(V)$ and $\hat{f}(\langle x \rangle) = \langle y \rangle$. [*Hint*: Let $\{x_1, \cdots, x_k\}$ be a basis for U and consider the cases.

   (a) $\langle x, x_1, \cdots, x_k \rangle = \langle y, x_1, \cdots, x_k \rangle$ and
   (b) $\langle x, x_1, \cdots, x_k \rangle \neq \langle y, x_1, \cdots, x_k \rangle$.]

5. Prove Theorem 5. [*Hint*: Let $\hat{g}$ be the given projective map, with g linear, and let $\{x_0, \cdots, x_n\}$ be a basis for V. Show by induction that there exist perspectivities $\hat{g}_0, \cdots, \hat{g}_n$ such that for each m, $\hat{g}_m \cdots \hat{g}_0 \cdot \hat{g}$ has $\langle x_0, \cdots, x_m \rangle$ for its set of fixed points, using Exercise 4.]

6. Let $f:V \to V$ be semilinear. Show that $\hat{f} = \hat{g}$ for some linear map g if and only if the automorphism associated with f is an inner automorphism. (Compare with Exercise 1.)

7. Show that the set of projectivities of $\mathscr{P}$ onto $\mathscr{P}$ is a *normal subgroup* of the set of collineations.

8. Why must D be commutative in Theorem 7?

SECTION 4.8

CROSS RATIO

You recall that in Chapter 2, Section 2.3 we defined the ratio of three collinear affine points p, q, x denoted $\alpha = \alpha(p, q:x)$, by $p - x = \alpha(q - x)$, where $q \neq x$. If we write the barycentric coordinate representation of x in terms of p, q, $x = \mu p + \tau q$ where $\mu + \tau = 1$, $\mu \neq 0$, then $\alpha(p, q:x) = \mu^{-1}\tau$. The objective of this section is to develop an analogue of the affine ratio in projective geometry. In affine geometry we were able to coordinatize points of a line in terms of two fixed points on it, but an analysis of the coordinatization of a projective space in Section 4.7 shows that three points are needed on a projective line. (Given a point P, a hyperplane H, $P \notin H$, and a line L, $P \in L$, L is coordinatized by the dilatations about P, that is, in terms of P, $L \cap H$ and some fixed point on L.)

We shall, throughout this section, restrict our attention to $\mathscr{P}(D^{n+1})$ where $n \geq 2$ and D is a field, $D \neq \mathbb{Z}_2$.

Proposition 1

Let $P_1, P_2, P_3, P_4 \in \mathscr{P}(D^{n+1})$ be distinct and collinear, and let $P_i = \langle x_i \rangle = \langle y_i \rangle$, $i = 1, 2, 3, 4$. Then there exist unique $\mu_1, \mu_2, \tau_1, \tau_2, \theta_1, \theta_2, \rho_1, \rho_2 \in D$ such that $x_3 = \mu_1 x_1 + \mu_2 x_2$, $x_4 = \tau_1 x_1 + \tau_2 x_2$, $y_3 = \theta_1 y_1 + \theta_2 y_2$, and $y_4 = \rho_1 y_1 + \rho_2 y_2$. Moreover

$$(\tau_1{}^{-1}\tau_2)^{-1}(\mu_1{}^{-1}\mu_2) = (\rho_1{}^{-1}\rho_2)^{-1}(\theta_1{}^{-1}\theta_2).$$

Proof

The existence of the scalars giving the stated linear combinations is immediate, for $\{x_i, y_i\}$ lies in a 2-dimensional subspace by the collinearity of the P_i and the distinctness of the P_i makes $\{x_1, x_2\}$ and $\{y_1, y_2\}$ bases of this subspace.

In order to establish the final equation of the conclusion we note that for each i there is an $\alpha_i \neq 0$ such that $y_i = \alpha_i x_i$, and that the distinctness of P_1, P_2, P_3, P_4 implies that none of $\mu_1, \mu_2, \tau_1, \tau_2, \theta_1, \theta_2, \rho_1, \rho_2$ are 0. Then

$$\begin{aligned} y_3 = \alpha_3 x_3 &= \alpha_3 \mu_1 x_1 + \alpha_3 \mu_2 x_2 \\ &= \alpha_3 \mu_1 \alpha_1^{-1} y_1 + \alpha_3 \mu_2 \alpha_2^{-1} y_2, \end{aligned}$$

so $\theta_1 = \alpha_3 \mu_1 \alpha_1^{-1}$ and $\theta_2 = \alpha_3 \mu_2 \alpha_2^{-1}$. Similarly, $\rho_1 = \alpha_4 \tau_1 \alpha_1^{-1}$ and $\rho_2 = \alpha_4 \tau_2 \alpha_2^{-1}$, whence

$$\begin{aligned} (\rho_1^{-1}\rho_2)^{-1}(\theta_1^{-1}\theta_2) &= (\alpha_1 \tau_1^{-1} \alpha_4^{-1} \alpha_4 \tau_2 \alpha_2^{-1})^{-1}(\alpha_1 \mu_1^{-1} \alpha_3^{-1} \alpha_3 \mu_2 \alpha_2^{-1}) \\ &= \alpha_2 \tau_2^{-1} \tau_1 \mu_1^{-1} \mu_2 \alpha_2^{-1} \\ &= (\tau_1^{-1}\tau_2)^{-1}(\mu_1^{-1}\mu_2). \end{aligned}$$

DEFINITION 1 Let P_1, P_2, P_3, P_4 be distinct collinear points in $\mathscr{P}(D^{n+1})$, $P_i = \langle x_i \rangle$, $x_3 = \mu_1 x_1 + \mu_2 x_2$, and $x_4 = \tau_1 x_1 + \tau_2 x_2$. We define the *cross ratio of* P_1, P_2, P_3, P_4, denoted CR $(P_1, P_2 : P_3, P_4)$, by CR $(P_1, P_2 : P_3, P_4) = \tau_1 \tau_2^{-1} \mu_1^{-1} \mu_2$. Observe that the condition requiring P_1, P_2, P_3, P_4 to be distinct is more restrictive than is necessary. The only distinctness properties required are that $P_4 \neq P_1$, $P_2 \neq P_1$, and $P_3 \neq P_2$ (see Exercise 3).

Proposition 1 guarantees us that this definition is independent of the choice of representatives for P_i. We now investigate the behavior of the cross ratio in relation to collineations.

Proposition 2

Let $\varphi : \mathscr{P}(D^{n+1}) \to \mathscr{P}(D^{n+1})$ be a collineation. Then there exists an automorphism $\sigma : D \to D$ such that for all sets P_1, P_2, P_3, P_4 of distinct collinear points,

$$\text{CR}\,(P_1, P_2 : P_3, P_4)^\sigma = \text{CR}\,(\varphi(P_1), \varphi(P_2) : \varphi(P_3), \varphi(P_4)).$$

Proof

We know that there is a semilinear map $f : D^{n+1} \to D^{n+1}$ with automorphism $\sigma : D \to D$ such that $\varphi\langle x \rangle = \langle (x)f \rangle$ for all points in $\mathscr{P}(D^{n+1})$. Letting P_1, P_2, P_3, P_4 be distinct collinear points and $P_i = \langle x_i \rangle$ we have, as usual, $x_3 = \mu_1 x_1 + \mu_2 x_2$ and $x_4 = \tau_1 x_1 + \tau_2 x_2$. Then $(x_3)f = \mu_1^\sigma (x_1)f + \mu_2^\sigma (x_2)f$ and $(x_4)f = \tau_1^\sigma (x_1)f + \tau_2^\sigma (x_2)f$, and

$$\begin{aligned} \text{CR}\,(\varphi(P_1), \varphi(P_2) : \varphi(P_3), \varphi(P_4)) &= \text{CR}\,(\langle (x_1)f \rangle, \langle (x_2)f \rangle : \langle (x_3)f \rangle, \langle (x_4)f \rangle) \\ &= ((\tau_1^\sigma)^{-1} \tau_2^\sigma)^{-1}((\mu_1^\sigma)^{-1} \mu_2^\sigma) \\ &= ((\tau_1^{-1}\tau_2)^{-1}(\mu_1^{-1}\mu_2))^\sigma \\ &= \text{CR}\,(P_1, P_2 : P_3, P_4)^\sigma. \end{aligned}$$

Corollary

If φ is a projectivity,

$$\mathrm{CR}\,(P_1, P_2 : P_3, P_4) = \mathrm{CR}\,(\varphi(P_1), \varphi(P_2) : \varphi(P_3), \varphi(P_4)).$$

The following proposition tells us how the cross ratio behaves under permutations of points.

Proposition 3

Let P_1, P_2, P_3, P_4 be distinct collinear points. Then

(a) $\mathrm{CR}\,(P_1, P_2 : P_3, P_4) = \mathrm{CR}\,(P_3, P_4 : P_1, P_2)$,
(b) $\mathrm{CR}\,(P_1, P_2 : P_3, P_4) = [\mathrm{CR}\,(P_1, P_2 : P_4, P_3)]^{-1}$,
(c) $\mathrm{CR}\,(P_1, P_2 : P_3, P_4) = 1 - \mathrm{CR}\,(P_1, P_3 : P_2, P_4)$.

Proof

The proof is left as an exercise for the reader.

Proposition 4

Let L_1, L_2, L_3, L_4 be distinct coplanar lines in $\mathscr{P}(D^{n+1})$ intersecting at a point P. Then

(a) there exists a line L such that $L \cap L_i$ is a point other than P for $i = 1, 2, 3, 4$;
(b) given a second line L' such that $L' \cap L_i$ is a point other than P for all i, then

$$\mathrm{CR}\,(L \cap L_1, L \cap L_2 : L \cap L_3, L \cap L_4) = \mathrm{CR}\,(L' \cap L_1, L' \cap L_2 : L' \cap L_3, L' \cap L_4).$$

Proof (See Figure 4.21.)

(a) There exist points $P_1 \in L_1$, $P_1 \neq P$, and $P_2 \in L_2$, $P_2 \neq P$ by PS (3). $(P_1 + P_2) \cap L_3 = P_3$ and $(P_1 + P_2) \cap L_4 = P_4$ are points by PS (2).
(b) Given a second line L' such that $L' \cap L_i = P_i'$, there exists a hyperplane containing L_4 and not P_1. Let $\varphi : \mathscr{P}(D^{n+1}) \to \mathscr{P}(D^{n+1})$ be the perspectivity leaving that hyperplane fixed and moving P_1 to P_1'. Also there exists a hyperplane containing L' and not P_4. Let φ' be the perspectivity leaving that hyperplane fixed and mapping P_4 to P_4'. Then $\varphi'\varphi(P_i) = P_i'$ for $i = 1, 2, 3, 4$, and so two applications of the corollary to Proposition 3 complete the proof.

DEFINITION 2 If L_1, L_2, L_3, L_4 are distinct coplanar lines in $\mathscr{P}(D^{n+1})$ intersecting at a point P, and if L is a line such that $L \cap L_i$ is a point other than P for all i, we define

$$\mathrm{CR}\,(L_1, L_2 : L_3, L_4) = \mathrm{CR}\,(L \cap L_1, L \cap L_2 : L \cap L_3, L \cap L_4).$$

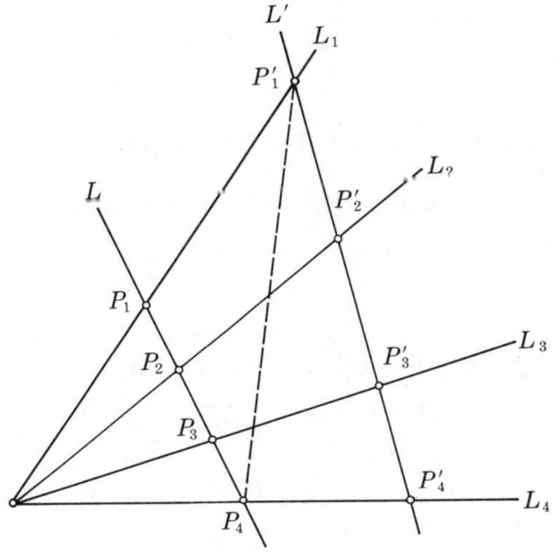

Figure 4.21

DEFINITION 3 The distinct collinear projective points P_1, P_2, P_3, P_4 are said to be *harmonic* if $\mathrm{CR}\,(P_1, P_2 : P_3, P_4) = -1$.

Proposition 4 tells us that the property of being harmonic is a function of the order in which the points are given. Given P_1, P_2, P_3, one can (and you will in Exercise 3) compute algebraically a point P_4 such that $\{P_1, P_2, P_3, P_4\}$ is harmonic. We shall, however, give a construction technique for finding P_4.

Given P_1, P_2, P_3 distinct collinear projective points, let Q be an arbitrary point not on $P_1 + P_2$ and Q' a point, $Q' \in P_1 + Q$, $Q' \neq P_1$, $Q' \neq Q$. (See Figure 4.22.) Let $R = (P_2 + Q') \cap (P_3 + Q)$, $R' = (P_1 + R) \cap (Q + P_2)$, $P_4 = (Q' + R') \cap (P_1 + P_2)$, and $S = (Q + R) \cap (Q' + R')$. We know that S, R, R', and P_4' are points because all the lines involved are coplanar. Therefore

$$\begin{aligned}
\mathrm{CR}\,(P_1, P_2 : P_3, P_4) &= \mathrm{CR}\,(Q + P_1, Q + P_2 : Q + P_3, Q + P_4)\\
&= \mathrm{CR}\,(Q + Q', Q + R' : Q + S, Q + P_4)\\
&= \mathrm{CR}\,(Q', R' : S, P_4)\\
&= \mathrm{CR}\,(R + Q', R + R' : R + S, R + P_4)\\
&= \mathrm{CR}\,(R + P_2, R + P_1 : R + P_3, R + P_4)\\
&= \mathrm{CR}\,(P_2, P_1 : P_3, P_4)\\
&= \mathrm{CR}\,(P_3, P_4 : P_2, P_1)\\
&= [\mathrm{CR}\,(P_3, P_4 : P_1, P_2)]^{-1}\\
&= [\mathrm{CR}\,(P_1, P_2 : P_3, P_4)]^{-1}.
\end{aligned}$$

Therefore $\mathrm{CR}\,(P_1, P_2 : P_3, P_4) = \pm 1$. But $\mathrm{CR}\,(P_1, P_2 : P_3, P_4) = 1 \Rightarrow P_3 = P_4$, which cannot be from our construction (why?). Therefore P_1, P_2, P_3, P_4 are harmonic. That P_4 is unique is demonstrated in Exercise 3.

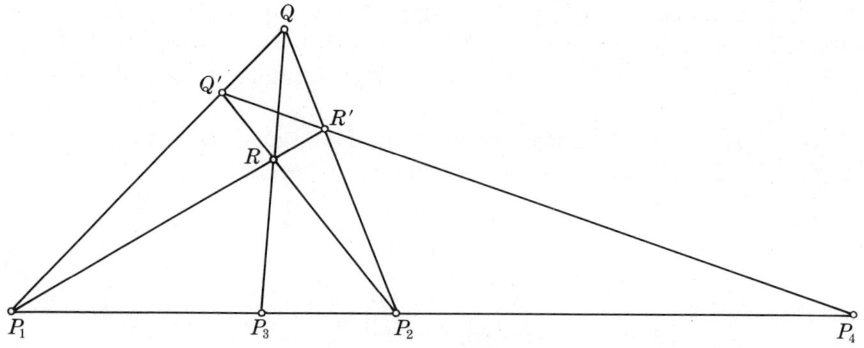

Figure 4.22

A careful reading of this construction yields the following.

Proposition 5

Let (P_1, P_2, P_3) be a triangle in $\mathscr{P}(D^{n+1})$ and choose $A_1, A_2, A_3, B_1, B_2, B_3 \in \mathscr{P}(D^{n+1})$ such that $A_1, B_1 \in P_2 + P_3$, $A_2, B_2 \in P_1 + P_3$, and $A_3, B_3 \in P_1 + P_2$, such that all nine points are distinct and such that the sets P_2, P_3, A_1, B_1; P_3, P_1, A_2, B_2; and P_1, P_2, A_3, B_3 are harmonic. Then $(P_1 + A_1)$,$(P_2 + A_2)$, $(P_3 + A_3)$ intersect at a point $\Leftrightarrow$ B_1, B_2, B_3 are collinear.

Proof

$\boxed{\Rightarrow}$: Let $S = (P_1 + A_1) \cap (P_2 + A_2) \cap (P_3 + A_3)$. (See Figure 4.23.) Then the triangles (P_1, P_2, P_3) and (A_1, A_2, A_3) are point perspective at S, so by Desargues' theorem the points $(A_1 + A_2) \cap (P_1 + P_2)$, $(A_2 + A_3) \cap (P_2 + P_3)$, and $(A_1 + A_3) \cap (P_1 + P_3)$ are collinear. But these are, by the

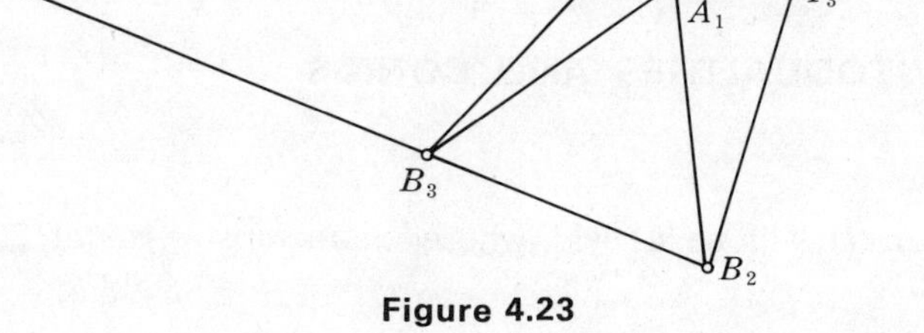

Figure 4.23

previous construction, B_3, B_1, and B_2, respectively. For example, to construct B_1 we choose an arbitrary point, say P_1, not on $P_2 + P_3$, a point on $P_1 + P_2$, say A_3, and obtain B_1 as $(A_2 + A_3) \cap (P_2 + P_3)$. A similar argument holds for B_2 and B_3.

$\boxed{\Leftarrow}$: B_1, B_2, B_3 collinear implies, by Desargues' theorem, that the triangles (P_1, P_2, P_3), (A_1, A_2, A_3) are point perspective.

EXERCISES

1. Prove Proposition 3.

2. Let $\tilde{\mathbb{R}} = \{\alpha \in \mathbb{R} : \alpha \neq 0, \alpha \neq 1\}$. Part (a) of Proposition 3 says that a certain permutation of P_1, P_2, P_3, P_4 is associated with the function $\rho_1 : \tilde{\mathbb{R}} \to \tilde{\mathbb{R}}$ given by $\rho_1(\alpha) = \alpha$. Part (b) says that another permutation is associated with the function $\rho_2 : \tilde{\mathbb{R}} \to \tilde{\mathbb{R}}$ given by $\rho_2(\alpha) = \alpha^{-1}$. Define ρ_3, ρ_4, ρ_5, $\rho_6 : \tilde{\mathbb{R}} \to \tilde{\mathbb{R}}$ by $\rho_3(\alpha) = 1 - \alpha$, $\rho_4(\alpha) = 1/(1 - \alpha)$, $\rho_5(\alpha) = (\alpha - 1)/\alpha$, and $\rho_6(\alpha) = \alpha/(\alpha - 1)$. Show that each of the twenty-four permutations of P_1, P_2, P_3, P_4 is associated with one of the functions $\rho_1, \cdots, \rho_6$. Show that this set of six functions is a group under composition. This group is often referred to as the *cross-ratio group*.

3. Let P_1, P_2, P_3 be collinear points with $P_2 \neq P_1$ and $P_3 \neq P_2$, and let $\alpha \in D$. Show that there exists a point $P_4 \in P_1 + P_2$ such that $\mathrm{CR}(P_1, P_2 : P_3, P_4) = \alpha$. Is P_4 unique?

4. Using Exercise 3 and Proposition 2 show that in $\mathscr{P}(\mathbb{R}^{n+1})$ the cross ratio of four collinear points is independent of the coordinatization, but that in $\mathscr{P}(\mathbb{C}^{n+1})$ this is not true. Show that for any field D the fact that P_1, P_2, P_3, P_4 are harmonic in $\mathscr{P}(D^{n+1})$ is independent of the choice of coordinates.

SECTION 4.9

AUTODUALITIES AND CONICS

In this section D will be a field whose characteristic is not 2 and $\mathscr{P} = \mathscr{P}(D^{n+1})$, $n \geq 2$.

DEFINITION 1 An *autoduality* of $\mathscr{P}$ is a map $\delta:\mathscr{F}(\mathscr{P}) \to \mathscr{F}(\mathscr{P})$ satisfying:

(**a**) $\delta^2 = 1$;
(**b**) if $A, B \in \mathscr{F}(\mathscr{P})$ and $A \subseteq B$ then $\delta(A) \supseteq \delta(B)$.

This is, of course, merely a special case of the duality theory discussed in Section 4.4 where $\mathscr{P}$ serves as its own dual. In particular the image of a k-dimensional flat will be an $(n - k - 1)$-dimensional flat. In the case of a projective plane these maps have traditionally been called *coreflections*. The main result we will prove is that all autodualities can be realized via σ-bilinear forms. As a preliminary we prove:

Lemma 1

Let $\sigma:D \to D$ be an automorphism and $\Gamma:D^{n+1} \times D^{n+1} \to D$ a σ-bilinear form, such that $\Gamma(x, y) = 0 \Rightarrow \Gamma(y, x) = 0$. Then the map $\delta:\mathscr{F}(\mathscr{P}) \to \mathscr{F}(\mathscr{P})$ defined by

$$\delta(A) = \{\langle y \rangle \in \mathscr{P}:\Gamma(x, y) = 0, \forall \langle x \rangle \in A\}$$

is an autoduality.

Proof

That $A \subseteq B$ implies $\delta(A) \supseteq \delta(B)$ is immediate. That $\delta^2(A) = A$ follows from the fact that $\dim(\delta(A)) = n + 1 - \dim A$ and hence $\dim(\delta^2(A)) = \dim(A)$. Since $A \subseteq \delta^2(A)$ we have $A = \delta^2(A)$. The dimension equation arises as follows.

Let $u \in D^{n+1}$. Then as in the proof of Theorem 2 of Chapter 3, Section 3.2, $K_u = \{v \in D^{n+1}:\Gamma(v, u) = 0\}$ is a subspace of dimension n. Let $x_1, \cdots, x_k$ be a basis of A. $\delta(A) = \bigcap_{i=1}^{k} K_{x_i}$, as in the proof of Theorem 2; the independence of the $\{x_i\}$ makes the $\{K_{x_i}\}$ distinct; and upon application of part (d), Exercise 6 of Chapter 2, Section 2.1, the result follows.

Theorem 1

Let $\delta:\mathscr{F}(\mathscr{P}) \to \mathscr{F}(\mathscr{P})$ be an autoduality. Then there exists a nonsingular σ-bilinear form $\Gamma:D^{n+1} \times D^{n+1} \to D$ such that

$$\delta(A) = \{\langle y \rangle \in \mathscr{P}:\Gamma(x, y) = 0, \forall \langle x \rangle \in A\}.$$

(and $\Gamma(x, y) = 0 \Rightarrow \Gamma(y, x) = 0, \forall x, y \in D^{n+1}$).

Proof

Let $\xi:\mathscr{F}(\mathscr{P}) \to \mathscr{F}(\mathscr{P})$ be the autoduality corresponding, as in Lemma 1, with the bilinear form whose matrix, relative to the standard basis of D^{n+1}, is the identity matrix, that is,

$$\xi(A) = \left\{ \langle(\beta_1, \cdots, \beta_{n+1})\rangle \in \mathscr{P} : (\alpha_1 \cdots \alpha_{n+1}) \begin{pmatrix} \beta_1 \\ \vdots \\ \beta_{n+1} \end{pmatrix} = 0, \right.$$

$$\left. \forall \langle(\alpha_1, \cdots, \alpha_{n+1})\rangle \in A \right\}.$$

Then by Proposition 5(c), Section 4.4, $\xi\delta^{-1}:\mathscr{P} \to \mathscr{P}$ is a collineation and so there exists by Theorem 3, Section 4.7 a semilinear transformation $f:D^{n+1} \to D^{n+1}$ with automorphism σ such that $\xi\delta^{-1} = \hat{f}$. If $\{e_1, \cdots, e_{n+1}\}$ is the standard basis of D^{n+1} and $(e_i)f = (\gamma_{i1}, \cdots, \gamma_{in+1})$ then

$$(\beta_1, \cdots, \beta_{n+1})f = (\beta_1{}^\sigma \cdots \beta_{n+1}{}^\sigma)(\gamma_{ij}).$$

We then have

$$\begin{aligned} \langle(\beta_1, \cdots, \beta_{n+1})\rangle \in \delta(A) &\Leftrightarrow \hat{f}\langle(\beta_1, \cdots, \beta_{n+1})\rangle \in \xi(A) \\ &\Leftrightarrow \langle(\beta_1{}^\sigma \cdots \beta_{n+1}{}^\sigma)(\gamma_{nj})\rangle \in \xi(A) \\ &\Leftrightarrow (\alpha_1 \cdots \alpha_{n+1})(\gamma_{ij})^T \begin{pmatrix} \beta_1{}^\sigma \\ \vdots \\ \beta_{n+1}{}^\sigma \end{pmatrix} \\ &= 0, \forall \langle(\alpha_1, \cdots, \alpha_{n+1})\rangle \in A. \end{aligned}$$

If we let $\Gamma:D^{n+1} \times D^{n+1} \to D$ be the σ-bilinear form whose matrix, relative to the standard basis, is $(\gamma_{ij})^T$, the result follows. The nonsingularity of Γ follows from the fact that f is 1-1 and onto and $\Gamma(x, y) = 0 \Rightarrow \Gamma(y, x) = 0$ follows from $\delta^2(A) = A$.

We turn now to the study of projective conics, which are a special case of coreflections in a plane. From this point on we will let $\mathscr{P} = \mathscr{P}(D^3)$.

DEFINITION 2 A *conic* C is a *nonempty* set of points of $\mathscr{P}$ such that for some symmetric bilinear form $\Gamma:D^3 \times D^3 \to D$, $C = \{\langle x\rangle \in \mathscr{P}:\Gamma(x, x) = 0\}$.

By Theorem 1 of Chapter 3, Section 3.4, we know that there is a homogenous polynomial of degree 2, $\lambda_1 X^2 + \lambda_2 Y^2 + \lambda_3 Z^2 + \lambda_4 XY + \lambda_5 XZ + \lambda_6 YZ$, and that C is the set of all points $\langle(\alpha_1, \alpha_2, \alpha_3)\rangle$ such that $(\alpha_1, \alpha_2, \alpha_3)$ is a nontrivial zero of the polynomial. We will refer to the equation obtained by setting the polynomial equal to 0 as the *equation of C*.

Proposition 1

The image of a conic under a collineation is a conic.

Proof

Let $\{e_1, e_2, e_3\}$ be the standard basis of D^3, C a conic relative to the symmetric bilinear form $\Gamma: D^3 \times D^3 \to D$ and let $(\gamma_{ij}) \in D_3$ be the matrix of Γ relative to the standard basis. A collineation is simply $\hat{f}$ for some semilinear transformation f with automorphism σ. If $(e_i)f = (\mu_{i1}, \mu_{i2}, \mu_{i3})$ for $i = 1, 2, 3$ then $(\mu_{ij}) \in D_3$ is a nonsingular matrix. Take $\Gamma': D^3 \times D^3 \to D$ to be the symmetric bilinear form whose matrix relative to the standard basis is $(\mu_{ij})^{-1}(\gamma_{ij})^\sigma ((\mu_{ij})^{-1})^T$ (this is symmetric by Exercise 5 of Chapter 3, Section 3.3, and the observation that (γ_{ij}) symmetric implies $(\gamma_{ij})^\sigma$ is symmetric).

Then

$$\Gamma'((\alpha_1, \alpha_2, \alpha_3)f, (\alpha_1, \alpha_2, \alpha_3)f)$$
$$= (\alpha_1{}^\sigma \alpha_2{}^\sigma \alpha_3{}^\sigma)(\mu_{ij})(\mu_{ij})^{-1}(\gamma_{ij})^\sigma((\mu_{ij})^{-1})^T(\mu_{ij})^T \begin{pmatrix} \alpha_1{}^\sigma \\ \alpha_2{}^\sigma \\ \alpha_3{}^\sigma \end{pmatrix}$$
$$= \left[(\alpha_1 \alpha_2 \alpha_3)(\gamma_{ij}) \begin{pmatrix} \alpha_1 \\ \alpha_2 \\ \alpha_3 \end{pmatrix} \right]^\sigma,$$

so $\Gamma'((x)f, (x)f) = 0 \Leftrightarrow \Gamma(x, x) = 0$.

A conic is said to be *nondegenerate* if its associated symmetric bilinear form is nonsingular and *degenerate* otherwise. We shall restrict our discussion to the analysis of nondegenerate conics. The subject of degenerate conics will be dealt with in the exercises. We shall call two conics *projectively equivalent* if there is a collineation taking one to the other. The first major result about nondegenerate conics is, as you might guess from the discussion in Section 4.1, that any two nondegenerate conics are projectively equivalent. To prove this we need the following preliminary results. Throughout this discussion we will be working with a nondegenerate conic C relative to the symmetric bilinear form Γ.

Proposition 2

C contains at most two points of a line L.

Proof

Observing that the image of a nondegenerate conic under a collineation is a nondegenerate conic and that a conic contains three collinear points if and only if all its images under collineations do, we may assume C contains the collinear points $\langle(1, 0, 0)\rangle$, $\langle(0, 1, 0)\rangle$, $\langle(\alpha, \beta, 0)\rangle$, $\alpha\beta \neq 0$.

If the equation of C is

$$\lambda_1 X^2 + \lambda_2 Y^2 + \lambda_3 Z^2 + \lambda_4 XY + \lambda_5 XZ + \lambda_6 YZ = 0,$$

then $\langle(1, 0, 0)\rangle$ on C yields $\lambda_1 = 0$, $\langle(0, 1, 0)\rangle$ on C yields $\lambda_2 = 0$, and $\langle(\alpha, \beta, 0)\rangle$ on C yields $\lambda_4 = 0$, so the equation becomes

$$\lambda_3 Z^2 + \lambda_5 XZ + \lambda_6 YZ = 0.$$

The matrix of the associated bilinear form, relative to the standard basis, is then

$$\begin{pmatrix} 0 & 0 & \frac{\lambda_5}{2} \\ 0 & 0 & \frac{\lambda_6}{2} \\ \frac{\lambda_5}{2} & \frac{\lambda_6}{2} & \lambda_3 \end{pmatrix},$$

which is patently singular, contradicting the assumption that C is nondegenerate. Therefore C does not contain three collinear points.

Proposition 3

Given any line L and any nondegenerate conic C, there is a 1-1 correspondence between the points of L and the points of C.

Proof

(See Figure 4.24.) Assume that there is a point $\langle x\rangle$ on C but not on L (we shall prove this in a moment). We claim the map

$$\langle y\rangle \to \langle -\tfrac{1}{2}\Gamma(y, y)x + \Gamma(x, y)y\rangle \in \langle x\rangle + \langle y\rangle$$

is a 1-1 map of L onto C.

It is a map into C because

$$\begin{aligned}\Gamma(-\tfrac{1}{2}\Gamma(y, y)x + \Gamma(x, y)y,\ -\tfrac{1}{2}\Gamma(y, y)x + \Gamma(x, y)y) \\ = \tfrac{1}{4}\Gamma(y, y)\Gamma(y, y)\Gamma(x, x) - \Gamma(y, y)\Gamma(x, y)\Gamma(x, y) + \Gamma(x, y)\Gamma(x, y)\Gamma(y, y) \\ = 0.\end{aligned}$$

Also $\{\langle y\rangle : \Gamma(x, y) = 0\}$ is a line, since Γ determines an autoduality, and intersects L in one point. The image of this point is $\langle x\rangle$. Given any other $\langle y_1\rangle \neq \langle y_2\rangle$ on L, the lines $\langle x\rangle + \langle y_1\rangle$ and $\langle x\rangle + \langle y_2\rangle$ are distinct and therefore intersect only at $\langle x\rangle$, which says that the map is 1-1. The map is onto for $\langle x\rangle$ is the image of $\{\langle y\rangle : \Gamma(x, y) = 0\} \cap L$, and any point $\langle u\rangle \in C$, $\langle u\rangle \neq \langle x\rangle$ is the image of $(\langle u\rangle + \langle x\rangle) \cap L$.

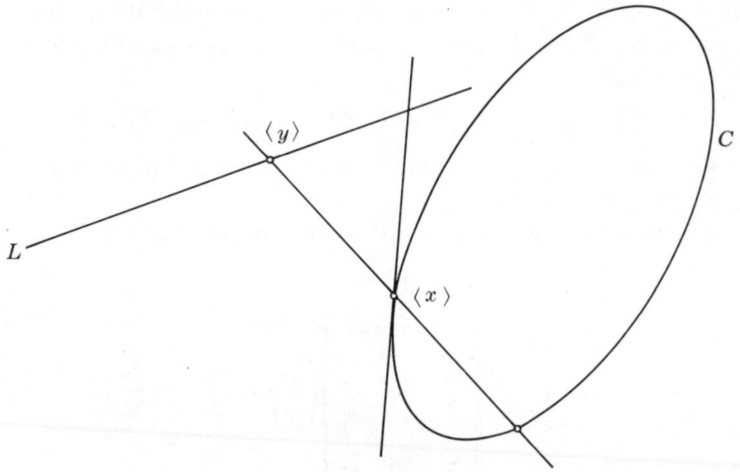

Figure 4.24

To eliminate the assumption at the beginning, C always contains some point $\langle x' \rangle$ and there is a line L' not containing $\langle x' \rangle$. By what we have done, C has as many points as L', that is, at least three, and so at least one of them is not on L.

If $\langle x \rangle \in C$ the line $\{\langle y \rangle \in \mathscr{P} : \Gamma(x, y) = 0\}$ is called the *tangent line to C at* $\langle x \rangle$. The tangent line intersects C only at $\langle x \rangle$, for if $\Gamma(x, y) = 0$ and $\Gamma(y, y) = 0$ then

$$\Gamma(x + y, x + y) = \Gamma(x, x) + 2\Gamma(x, y) + \Gamma(y, y) = 0,$$

which would imply C has three collinear points. In Exercise 2 you will show that given any line there is a linear equation

$$\alpha X + \beta Y + \gamma Z = 0$$

such that the line consists of points whose representatives are solutions and that conversely any linear equation determines a line.

Theorem 2

Any two nondegenerate conics are projectively equivalent.

Proof

We will show that any nondegenerate conic is projectively equivalent to the conic whose equation is $Y^2 - XZ = 0$.

If C is a nondegenerate conic it contains three noncollinear points P_1, P_2, P_3. Let P be the point of intersection of the tangents to C at P_1 and

P_2. No three of P, P_1, P_2, P_3 are collinear and so there is a projective map of $\mathscr{P}$ to $\mathscr{P}$ taking P to $\langle(0, 1, 0)\rangle$, P_1 to $\langle(1, 0, 0)\rangle$, P_2 to $\langle(0, 0, 1)\rangle$, and P_3 to $\langle(1, 1, 1)\rangle$. Let

$$\lambda_1 X^2 + \lambda_2 Y^2 + \lambda_3 Z^2 + \lambda_4 XY + \lambda_5 XZ + \lambda_6 YZ = 0$$

be the equation of the image conic. Since $(1, 0, 0)$ and $(0, 0, 1)$ are roots of the equation, $\lambda_1 = \lambda_3 = 0$.

The equation of the tangent to the conic at $\langle(1, 0, 0)\rangle$ is

$$0 = (1 \quad 0 \quad 0)\begin{pmatrix} 0 & \frac{\lambda_4}{2} & \frac{\lambda_5}{2} \\ \frac{\lambda_4}{2} & \lambda_2 & \frac{\lambda_6}{2} \\ \frac{\lambda_5}{2} & \frac{\lambda_6}{2} & 0 \end{pmatrix}\begin{pmatrix} X \\ Y \\ Z \end{pmatrix} = \frac{\lambda_4}{2} Y + \frac{\lambda_5}{2} Z,$$

and the equation of the tangent at $\langle(0, 0, 1)\rangle$ is $0 = (\lambda_5/2)X + (\lambda_6/2)Y$. Since $\langle(0, 1, 0)\rangle$ is on both tangents we have $\lambda_4 = \lambda_6 = 0$.

Finally $\langle(1, 1, 1)\rangle$ on the conic yields $\lambda_2 + \lambda_5 = 0$ and the equation of the conic is $Y^2 - XZ = 0$.

We now have the tools in hand to develop most of the classical theory of projective conics. We shall close this section and the book with a few of the basic results.

Theorem 3 (Steiner)

Let C be a conic and P_1, P_2, P_3, P_4 four distinct points on C. If P, P' are any points on C and $L_i = P + P_i$, $L_i' = P' + P_i$, then

$$CR(L_1, L_2 : L_3, L_4) = CR(L_1', L_2' : L_3', L_4')$$

whenever both sides are defined.

Proof

There exists a projective coordinatization under which P_1, P_2, P_3, P_4 are respectively $\langle(1, 0, 0,)\rangle$, $\langle(0, 1, 0)\rangle$, $\langle(0, 0, 1)\rangle$, and $\langle(1, 1, 1)\rangle$. Substitution of these points into the general equation of a conic yields

$$X(Y - Z) - \mu Y(X - Z) = 0$$

for some $\mu \in D$, $\mu \neq 0$ as the equation of the conic.

Let $P = \langle(\alpha, \beta, \gamma)\rangle$ be on the conic. Elementary calculations yield

$$(P + P_3) \cap (P_1 + P_2) = \langle(\alpha, \beta, 0)\rangle$$

and $$(P + P_4) \cap (P_1 + P_2) = \langle(\alpha - \gamma, \beta - \gamma, 0)\rangle.$$

If any of α, β, $\alpha - \gamma$, or $\beta - \gamma$ are 0 the conic contains three collinear points contradicting the assumption that it is nonsingular, and therefore

$$CR\,(L_1, L_2 : L_3, L_4) = \frac{\alpha}{\beta}\,\frac{\beta - \gamma}{\alpha - \gamma} = \frac{\mu\beta(\alpha - \gamma)}{\beta(\alpha - \gamma)} = \mu.$$

Clearly the cross ratio is independent of the choice of P.

If L is a line, it has (see Exercise 2) an equation $\tilde{L}(X, Y, Z) = \alpha X + \beta Y + \gamma Z = 0$, which we will abbreviate to $\tilde{L} = 0$. We shall denote the fact that $\langle x\rangle \in L$ by writing $\tilde{L}(x) = 0$. Let $\langle x_1\rangle$, $\langle x_2\rangle$, $\langle x_3\rangle$, and $\langle x_4\rangle$ be four distinct points, $L_1 = \langle x_1\rangle + \langle x_2\rangle$, $L_2 = \langle x_2\rangle + \langle x_3\rangle$, $L_3 = \langle x_3\rangle + \langle x_4\rangle$, and $L_4 = \langle x_4\rangle + \langle x_1\rangle$. Then for any $\alpha \in D$,

$$\tilde{L}_1\tilde{L}_3 + \alpha\tilde{L}_2\tilde{L}_4 = 0,$$

where

$$(\tilde{L}_1\tilde{L}_3 + \alpha\tilde{L}_2\tilde{L}_4)(X, Y, Z) = \tilde{L}_1(X, Y, Z)\tilde{L}_3(X, Y, Z) + \alpha\tilde{L}_2(X, Y, Z)\tilde{L}_4(X, Y, Z)$$

is the equation of a conic since it is a homogenous polynomial of degree 2, and it is easily seen to contain $\langle x_1\rangle$, $\langle x_2\rangle$, $\langle x_3\rangle$, and $\langle x_4\rangle$.

We utilize these observàtions in the proof of:

Theorem 4 (Pascal)

Let C be a nondegenerate conic containing the distinct points $\langle x_i\rangle$, $i = 1, \cdots, 6$. Let

$$\langle p_1\rangle = (\langle x_1\rangle + \langle x_2\rangle) \cap (\langle x_4\rangle + \langle x_5\rangle),$$
$$\langle p_2\rangle = (\langle x_2\rangle + \langle x_3\rangle) \cap (\langle x_5\rangle + \langle x_6\rangle),$$

and

$$\langle p_3\rangle = (\langle x_3\rangle + \langle x_4\rangle) \cap (\langle x_6\rangle + \langle x_1\rangle).$$

Then $\langle p_1\rangle$, $\langle p_2\rangle$, and $\langle p_3\rangle$ are collinear.

Proof

Let $L_1 = \langle x_1\rangle + \langle x_2\rangle$, $L_2 = \langle x_2\rangle + \langle x_3\rangle$, $L_3 = \langle x_3\rangle + \langle x_4\rangle$, $L_1' = \langle x_4\rangle + \langle x_5\rangle$, $L_2' = \langle x_5\rangle + \langle x_6\rangle$, $L_3' = \langle x_6\rangle + \langle x_1\rangle$, and $L = \langle x_1\rangle + \langle x_4\rangle$. (See Figure 4.25.) By our previous remarks, for any $\alpha \in D$, $\tilde{L}_1\tilde{L}_3 - \alpha\tilde{L}_2\tilde{L} = 0$ is the equation of a conic containing $\langle x_1\rangle$, $\langle x_2\rangle$, $\langle x_3\rangle$, and $\langle x_4\rangle$. Since $\langle x_5\rangle \notin L$, $\langle x_5\rangle \notin L_i$, $i = 1, 2, 3$ (else C would contain three collinear points), if we take

$$\alpha = \frac{\tilde{L}_1(x_5)\tilde{L}_3(x_5)}{\tilde{L}_2(x_5)\tilde{L}(x_5)}$$

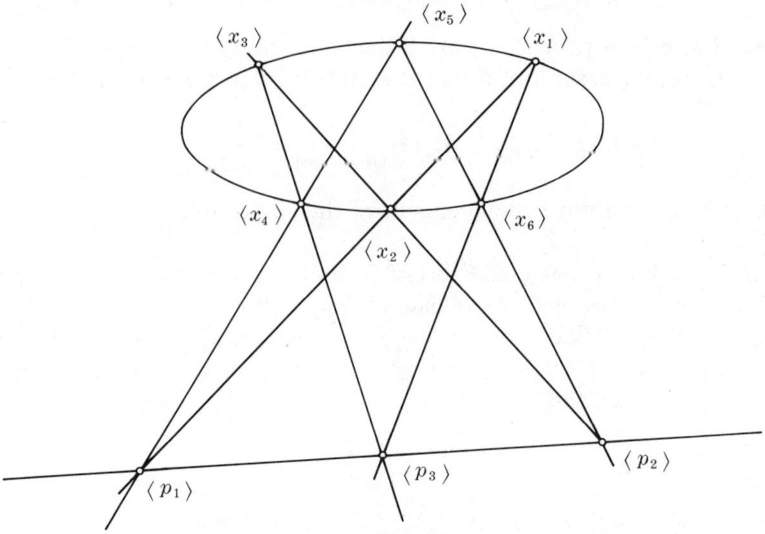

Figure 4.25

we must have the equation of C.

Similarly

$$\tilde{L}_1'\tilde{L}_3' - \beta\tilde{L}_2'\tilde{L} = 0,$$

where

$$\beta = \frac{\tilde{L}_1'(x_2)\tilde{L}_3'(x_2)}{\tilde{L}_2'(x_2)\tilde{L}(x_2)}$$

is the equation of C. Subtracting these yields the identity

$$\tilde{L}_1\tilde{L}_3 - \tilde{L}_1'\tilde{L}_3' = \tilde{L}(\alpha\tilde{L}_2 - \beta\tilde{L}_2'),$$

whence the conics $\tilde{L}_1\tilde{L}_3 - \tilde{L}_1'\tilde{L}_3' = 0$ and $\tilde{L}(\alpha\tilde{L}_2 - \beta\tilde{L}_2') = 0$ are the same.

If $\tilde{L}(p_1) = 0$, that is, $\langle p_1\rangle \in \langle x_1\rangle + \langle x_4\rangle$, then $\langle p_1\rangle \in \langle x_1\rangle + \langle x_2\rangle \Rightarrow \langle x_1\rangle, \langle x_2\rangle$, and $\langle x_4\rangle$ collinear which contradicts the nondegeneracy of C. Similarly $\tilde{L}(p_2) \neq 0$, $\tilde{L}(p_3) \neq 0$. Since $p_2 \in L_2$ and $p_2 \in L_2'$,

$$(\alpha\tilde{L}_2 - \beta\tilde{L}_2')(p_2) = \alpha\tilde{L}_2(p_2) - \beta\tilde{L}_2'(p_2) = 0.$$

Also

$$\begin{aligned}(\tilde{L}_1\tilde{L}_3 - \tilde{L}_1'\tilde{L}_3')(p_1) &= \tilde{L}_1(p_1)\tilde{L}_3(p_1) - \tilde{L}_1'(p_1)\tilde{L}_3'(p_1)\\ &= 0\tilde{L}_3(p_1) - 0\tilde{L}_3'(p_1)\\ &= 0,\end{aligned}$$

and

$$(\tilde{L}_1\tilde{L}_3 - \tilde{L}_1'\tilde{L}_3')(p_3) = 0.$$

Since $\tilde{L}(p_i) \neq 0$, $i = 1, 2, 3$, we have

$$(\alpha\tilde{L}_2 - \beta\tilde{L}_2')(p_i) = 0,\ i = 1, 2, 3,$$

that is, $\langle p_1 \rangle$, $\langle p_2 \rangle$, and $\langle p_3 \rangle$ are on the line whose equation is $\alpha\tilde{L}_2 - \beta\tilde{L}_2' = 0$.

Pascal's theorem can be reformulated as follows. A *simple hexagon* is a set of six distinct points, in some order, such that the six joins of successive points are all distinct lines. The theorem states that if the vertices of a simple hexagon lie on a nondegenerate conic then the intersections of the opposite sides are collinear.

EXERCISES

1. Let $P_1 = \langle(1, 0, 0)\rangle$, $P_2 = \langle(0, 0, 1)\rangle$, $P_3 = \langle(1, 1, 1)\rangle$, and $P_4 = \langle(1, -1, 1)\rangle \in \mathscr{P}(\mathbb{R}^3)$. Show that the set of all $P \in \mathscr{P}(\mathbb{R}^3)$ such that

$$\mathrm{CR}\,(P + P_1, P + P_2 : P + P_3, P + P_4) = 2$$

is a conic. From this you should be able to conjecture and prove a theorem.

2. Show that every line in $\mathscr{P}(\mathbb{R}^3)$ is the set of points whose representatives are solutions of the equation $\mu_1 X + \mu_2 Y + \mu_3 Z = 0$, for suitable μ_1, μ_2, μ_3. Show conversely that given any such equation there is a line whose points (or more properly their representatives) are solutions to the equation.

3. Classify the degenerate conics as follows. If C is a degenerate conic the matrix of its bilinear form Γ is of rank one or two. Show that if the matrix of Γ is of rank one the conic is simply a line squared. Show that if the matrix of Γ is of rank two then C is either a point or the product of two lines.

4. The conics we have discussed are called point conics. These have a dual called line conics, which come about as follows. If L is a line it has an equation $\mu_1 X + \mu_2 Y + \mu_3 Z = 0$ (see Exercise 2). But also $\alpha \cdot \mu_1 X + \alpha \cdot \mu_2 Y + \alpha \cdot \mu_3 Z = 0$ is an equation of the line for any $\alpha \neq 0$, so we may identify L with the "coordinates" $\langle \mu_1, \mu_2, \mu_3 \rangle$. If $(\gamma_{ij}) \in D_3$ is a symmetric matrix a *line conic* is

$$\left\{ \langle(\mu_1, \mu_2, \mu_3)\rangle : (\mu_1 \mu_2\, \mu_3)(\gamma_{ij}) \begin{pmatrix} \mu_1 \\ \mu_2 \\ \mu_3 \end{pmatrix} = 0 \right\},$$

provided this set of lines is nonempty. Now state and prove the dual to Pascal's theorem. This is known as Brianchon's theorem.

5. Let $P_1, P_2, P_3, P_4, P_5 \in \mathscr{P}(D^3)$ be five distinct points no three of which are collinear.

   (a) Show there is a unique conic containing all five points.
   (b) Show there is a unique conic containing P_1, P_2, P_3, P_4 and whose tangent at P_4 contains P_5.

6. Let $C \subseteq \mathscr{P}(\mathbb{R}^3)$ be a conic, and H a line. Consider the image of C in $\mathscr{P}(\mathbb{R}^3)/H$. Show that this image is an ellipse $\Leftrightarrow H \cap C = \varnothing$, is a parabola $\Leftrightarrow H \cap C$ is one point, and is a hyperbola $\Leftrightarrow H \cap C$ is two points. [*Hint*: Assume a coordinatization of $\mathscr{P}(\mathbb{R}^3)$ such that $\tilde{H}(X, Y, Z) = Z$, so that the points of $\mathscr{P}(\mathbb{R}^3)/H$ are $\{\langle(\alpha, \beta, 1)\rangle : \alpha, \beta \in \mathbb{R}\}$. Now find the form of the polynomial equation of $C \cap H$ as well as the form of $C \cap \mathscr{P}(\mathbb{R}^3)/H$.]

7. If H is a line in $\mathscr{P}(\mathbb{R}^3)$, then there is a 1-1 correspondence between points in $\mathscr{P}(\mathbb{R}^3)/H$ and affine points in the affine plane $\tilde{H}(X, Y, Z) = \beta$ for any $\beta \neq 0$ in $\mathbb{R}$. For example, if H is the line $\{\langle(\alpha, \beta, 0)\rangle : \alpha, \beta \in \mathbb{R}\}$, then the affine plane $\{(\alpha, \beta, 1) : \alpha, \beta \in \mathbb{R}\}$ would be appropriate. Give a pictorial discussion of Exercise 6 by considering a coordinatization that gives rise to the equation $X^2 + Y^2 - Z^2 = 0$ for the conic C. Now consider the effect of letting H be respectively $Z = 0$, $X = 0$, and $Y + Z = 0$.

Glossary of Symbols

SYMBOL	PAGE
$\sum_{i=1}^{n}$	20
$< S >$	20
$\setminus$	22
$\{e_1, \cdots, e_n\}$	24
$L_1 + \cdots + L_n$	24
$\dim (V)$	26
$\oplus$	27
$L_1 \oplus \cdots \oplus L_n$	28
$\mathbb{R}^n[x]$	29
Ker (f)	31
Im (f)	31
$\approx$	32
$GL(V)$	34
Aut (V)	34
$x + H$	35
V/H	38
I_n, I	46
$D_{n \times m}, D_n$	46
$m(f, B, C)$, $m(f, B)$	46
(α_{ij})	46
$\mathrm{Hom}_D(V, W)$	46
$m(x, B)$	50
$\mathscr{Q}$	52
e_{ij}	55
$\mathrm{E}(\lambda; i)$	55
$\mathrm{E}(\lambda; i, j)$	55
$E(i, j)$	55
A^T	60
V_D	65
V^*	65
V^{**}	66
Ann (H)	67
$g \cdot H$	69
$H \lhd G$	70
G/H	70
$\mathbb{Z}_p$	72
α^σ	72
$S - T$	74
$a_0 + U$	74
$L(A)$	74
$J(S)$	77
$J(x_1, \cdots, x_n)$	77
$\mathscr{L}(V)$	80
$\mathscr{A}(V)$	80
$A \parallel B$	83
$\alpha(p, q : x)$	86
$\triangle$	87
t_v	98
Coll (V)	104
$A(V)$	104
$T(V)$	104
$\lVert x \rVert$	110
$x \cdot y$	110
$\Gamma(x, y)$	111
$(\alpha_{ij})^\sigma$	112
$m(\Gamma, B)$	112
RK(Γ)	115
LK(Γ)	115
K_x	117
$S\perp$	119
$S \perp T$	119
$d(A)$	131
$d(f)$	131
adj (A)	147
V_λ	148
f^*	151
$\overline{V}$	161
AP(1), AP(2), AP(3)	161
PP(1), PP(2), PP(3)	161
$(\mathscr{P}, \Lambda)$	165
PS(1), PS(2)	165
$\mathscr{F}(\mathscr{P})$	166
$Z H_i$	166
$P + Q$	166
p-dim(A)	169
$\mathscr{P}/H$, Λ/H, A/H	172
F^*	177
$\delta(A)$	177
$\mathscr{P}^*, \Lambda^*$	178
Aut $(\mathscr{P})$	187
CR $(P_1, P_2 : \mathrm{P}_3, P_4)$	214
CR $(L_1, L_2 : L_3, L_4)$	215
$\tilde{L}$	225

Index

Index

A

B

C